D1601591

GROUPS, REPRESENTATIONS, AND CHARACTERS

GROUPS, REPRESENTATIONS, AND CHARACTERS

Victor E. Hill
Associate Professor of Mathematics
Williams College

HAFNER PRESS
A Division of Macmillan Publishing Co., Inc.
NEW YORK

Collier Macmillan Publishers
LONDON

Hafner Press
A Division of Macmillan Publishing Co., Inc.
866 Third Avenue, New York, N.Y. 10022

Collier Macmillan Canada, Ltd.

Library of Congress Catalog Card Number: 75–43362

Printed in the United States of America

printing number

1 2 3 4 5 6 7 8 9 10

Library of Congress Cataloging in Publication Data

Hill, Victor E
Groups, representations, and characters.

Bibliography: p.
Includes index.
1. Groups, Theory of. 2. Representations of groups.
3. Characters of groups. I. Title.
QA171.H64 1976 512'.2 75-43362
ISBN 0-02-846790-6

To Christi

Contents

Preface

This book is addressed to three potential audiences: (1) undergraduate majors or beginning graduate students in mathematics who wish to have a rapid survey of representation theory at a lower level of sophistication than that of the classic work by Curtis and Reiner; (2) students in chemistry, physics, or geology, who are likely to encounter groups and characters in such areas as crystallography and quantum mechanics; and (3) physical scientists whose experience with groups and characters, though extensive, has not been rigorous and who would like to have some sense of the mathematics lying behind the techniques used in applications.

All three constituencies have been represented in the one-semester course which I have taught at Williams College and from which this text is derived. The course prerequisite is the same as that for the book: a semester of college-level linear algebra and a basic familiarity with elementary logic and set theory (including the abbreviation *iff* for "if and only if"). Most of the needed results on matrices and vector spaces are recalled in the text, though only a few are proved here.

The text is intended to be a rapid survey, and therefore it emphasizes examples and applications of the theorems and avoids most of the longer and more difficult proofs. It also takes for granted basic properties of the natural numbers such as the Euclidean algorithm. Representations are limited to the real and complex numbers and, for the most part, to finite groups.

Some of the exercises stress computation of illustrative examples, others the development of the theory; the reader or instructor can make an appropriate choice among those provided. In a class it may prove desirable to incorporate some of the problems into the lecture. One should note, however, those exercises marked by a dot in the margin; they present results used in later sections and hence should be tried by each student or reader.

To avoid including extended expositions of ideas from chemistry and physics in a mathematics book, I have made the references to physical science applications primarily as indications of the directions in which they lie. Many of the books listed in the bibliography build on a greater scientific background than I have assumed here and provide detailed discussions of specific applications.

In the sections on representations and characters, I have adopted for the most part the notations of Curtis and Reiner: they are clear, and they facilitate the transition from this introduction to the comprehensive treatment in their work. If the reader who uses this book (as an undergraduate text, as supplementary material in an abstract algebra course, or as an exploration of the mathematical side of groups and characters) is thereby motivated to go on to advanced books on groups and representations, the purpose served will be not unlike that of a translation of a literary work that leads the reader to learn the language in order to read the original.

My first acknowledgment is to Professor Charles W. Curtis, who introduced me to representation theory, guided my graduate work, and has been a valued friend ever since. The idea of a group acting on a point set is due to Professor Helmut Wielandt. Four of my colleagues at Williams have provided generous help in the preparation of this book: I am indebted to James F. Skinner and Daniel A. Kleier for references to chemical applications, to William R. Moomaw for assistance with the material on space groups, and in particular to David A. Park for invaluable contributions to the section on applications to physics. My former students Craig S. Billie and David M. McCord read the manuscript and offered useful suggestions; my group theory class in the fall of 1975 used the manuscript as a text and made pertinent comments. Claude Conyers and David F. Biesel of Macmillan supplied many good editorial ideas. Final thanks are due to Eileen Sprague for her expert typing of the manuscript.

Part One GROUPS

Section 1
The Square-Bipyramid Group

Many approaches can be taken to defining a group. It is possible to abstract the concept from familiar properties of addition and multiplication of integers and real numbers. This definition is often taken in textbooks on abstract algebra. However, from the physical point of view it often proves helpful to start with a set of points in space and to consider their symmetries. This avenue is not far removed from the earliest study of groups, when Cauchy and Galois viewed a group as the transformations that can be made among the roots of a polynomial equation, subject to some very general conditions.

By a transformation of a set of points we mean an exchange (or *permutation*) in which each point goes to the place formerly occupied by some point of the set, and no two points go to the same location. A point may, of course, be left fixed by such an exchange. If α is one of our points and g is a transformation, we denote by α^g the point to whose original place α is carried by g.

For example, let Ω be the set consisting of the six vertices of a double pyramid over a square base, where the faces of the pyramid are isosceles but not equilateral triangles. Such a solid is sometimes called a *square bipyramid.* In figure 1, the four edges meeting at vertex 1 and the four meeting at vertex 6 are all equal, but are of a different length than the sides of the central square.* We denote by

* The square bipyramid arises in applications of group theory to chemistry, for example, in connection with the symmetry of the trans-dichlorotetramminecobalt (III) cation $Co(NH_3)_4Cl_2^+$. Neglecting the hydrogen atoms, which is usually done, the "local symmetry" of the cobalt is square bipyramidal with a Cl at vertices 1 and 6, with an NH_3 group at each of vertices 2 through 5, and with the Co at the center of the square base.

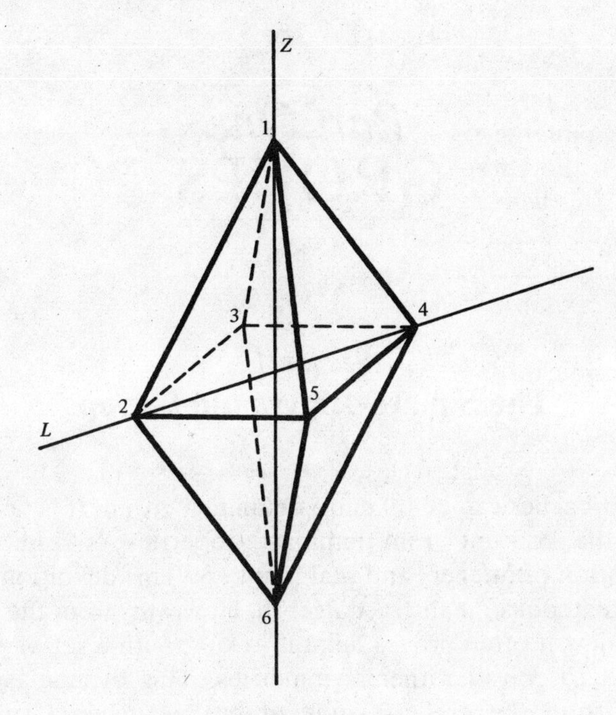

Figure 1

G the set of transformations that move the solid (which we may call a crystal) *rigidly* so that the six vertices are exchanged among themselves, but the solid is not disassembled or distorted. The lines L and Z are fixed, but the solid may be moved in space so that the outline shown in figure 1 is preserved, even if the numbered vertices fall in different places with respect to lines L and Z.

The first motion we consider from G is a rotation of 90° clockwise about line Z (viewed from above). If we denote this transformation by r, we have $2^r = 3$, meaning that vertex 2 moves into the space formerly occupied by vertex 3. Similarly, $3^r = 4$, $4^r = 5$, and $5^r = 2$. Since vertices 1 and 6 remain fixed under this rotation, we have $1^r = 1$ and $6^r = 6$.

If we repeat the motion r, we have effected a rotation through 180°, which may conveniently be designated as r^2. Then $1^{r^2} = 1$, $2^{r^2} = 4$, $3^{r^2} = 5$, $4^{r^2} = 2$, $5^{r^2} = 3$, $6^{r^2} = 6$.

Since the notation we have been using is cumbersome, we introduce the cyclic notation often found in work on permutation groups. The motion r may be denoted by (2345), where this expression means that 2 is carried to 3, 3 to 4, 4 to 5, and the 5 at the end is carried back to the 2 at the beginning. Such a notation is called a *cycle*. Similarly, we shall denote r^2 by the pair of cycles (24) (35), the distinct terms in parentheses indicating that vertices 2 and 4 are interchanged, as are 3 and 5. The vertices left fixed by a transformation are not usually included at all, but when we need to emphasize them we may write cycles containing only one point; thus r could be written as (1)(2345)(6) and r^2 as (1)(24)(35)(6).

Note that r could also be denoted by (3452), which would be a compact notation for $3^r = 4$, $4^r = 5$, $5^r = 2$, $2^r = 3$, merely changing the order in which the displacements under r are listed.

A rotation through 270° may be denoted by r^3 or by (2543), which may be checked from figure 1. Since we have written r^3, it should be true that $r^3 = r \cdot r^2$, in some sense of the multiplication symbol. In fact, if we juxtapose our expressions

$$r \cdot r^2 = (2345)(24)(35), \tag{1.1}$$

we may form the expression for r^3 without recourse to figure1. The procedure is as follows. Consider some vertex appearing in (1.1), say 2. Its first appearance is in the first cycle, reading from the left, where 2 goes to 3. We then look for the next cycle containing a 3, namely (35), where 3 goes to 5. We are now at the end of the expression (1.1), and we conclude that 2 goes to 5, that is, that r^3 begins (25 . . .). Repeating the process, 5 goes to 2 in the first cycle, and 2 goes to 4 in the second. Since no subsequent cycle contains a 4, the process stops, and we have $r^3 = (254\ldots)$. Two more applications of the procedure yield $r^3 = (2543)$.

Since rotation through 360°, denoted by r^4, brings us back to the starting position, it is the same effect as no motion at all. The act of doing nothing to the points is called the *identity transformation*, or simply the *identity*, and is denoted by e. Thus $r^4 = e$, and a check proceeding as described above will produce $r^4 = (1)(2)(3)(4)(5)(6)$. (See exercise 1.1.)

Thus far vertices 1 and 6 of the square bipyramid have remained fixed. They may be interchanged, however, by a rotation through 180° with line L as axis. If we denote this motion by c, we have $c = (16)(35)$. Clearly, $c^2 = e$.

Now we ask about the effect of rc, that is, motion r followed by motion c. We may calculate

$$rc = (2345)(16)(35) = (16)(25)(34) \tag{1.2}$$

by the procedure described above. Note that cr (c followed by r) is *not* the same as rc; in fact,

$$cr = (16)(35)(2345) = (16)(23)(45),$$

and computation will show that $r^3c = cr$. The multiplication we are using here is simply *composition*, or following one transformation by another. The fact that the order in which the two operations are performed makes a difference in the result is described by saying that the operation (of "multiplication") is *noncommutative.*

We can make a multiplication table, but because of the noncommutativity we need the convention that the left-hand factor corresponds to the row and the right-hand factor to the column in which an entry of the table is written. Not all of the entries will have to be calculated as at (1.2); for any motion x we have $xe = ex = x$, and some computations can be done visually, for example $r^2 \cdot rc = r^3c$, where we simply recall what r^2 and r mean in terms of rotations. One additional transformation will appear as we construct the table: $r^2c = (16)(24)$, but this one additional motion is the only other one that will be needed. The eight motions we have discussed are called the *rigid symmetries* or *rigid transformations* of the square bipyramid.

EXERCISES

NOTE. The exercises marked with a dot in the margin are those to which reference is made in the text of subsequent sections. The reader should try all such exercises, even though in some instances a solution is given when the result stated in the exercise is used.

1.1. Verify that $r^2 = r \cdot r$ and $r \cdot r^3 = r^4 = e$ by the method described at (1.1).

1.2. Verify that $r^3c = cr$.

1.3. Verify that $r^2c = cr^2$. Then observe that $r^2x = xr^2$ for every transformation x in G. (In this case, we say that r^2 is in the *center* of G.)

• *1.4.* Complete the table at the top of page 5 for the group of symmetries of the square bipyramid.

• *1.5.* The *trigonal bipyramid* is similar to the solid in figure 1 except that the square base common to the two pyramids is replaced by an equilateral triangle. The sides of this triangular base are of different length from the slanted edges of the crystal. Find the set G of rigid transformations of this figure by imitating the work done for the square bipyramid, and prepare a table like that in exercise 1.4. (The

	e	r	r^2	r^3	c	rc	r^2c	r^3c
e	e	r	r^2	r^3	c	rc	r^2c	r^3c
r	r	r^2	r^3	e	rc	r^2c	r^3c	c
r^2	r^2	r^3	e		r^2c	r^3c	c	
r^3	r^3	e			r^3c	c		
c	c	r^3c			e			
rc	rc				r			
r^2c	r^2c				r^2			
r^3c	r^3c				r^3			

Table for the Square-Bipyramid Group

trigonal bipyramid arises in chemistry in connection with the symmetry of the phosphorus pentachloride molecule PCl_5 in the gaseous state.)

1.6. The *regular tetrahedron* is shown in figure 2; all edges are equal in length, and the four faces are congruent equilateral triangles. Two of the rigid symmetries of the tetrahedon are the rotations $a = (234)$ and $b = (123)$. Compute ab by juxtaposing cycles, and verify geometrically that this transformation is a rotation through 180° about a line joining the midpoints of edges 1–2 and 3–4. (The tetrahedron figures in applications to chemistry, for example, as the model of the methane molecule CH_4.)

• *1.7.* Find all twelve rigid symmetries of the regular tetrahedron; this set forms the *tetrahedral group*, which we shall meet again in section 7.

Figure 2

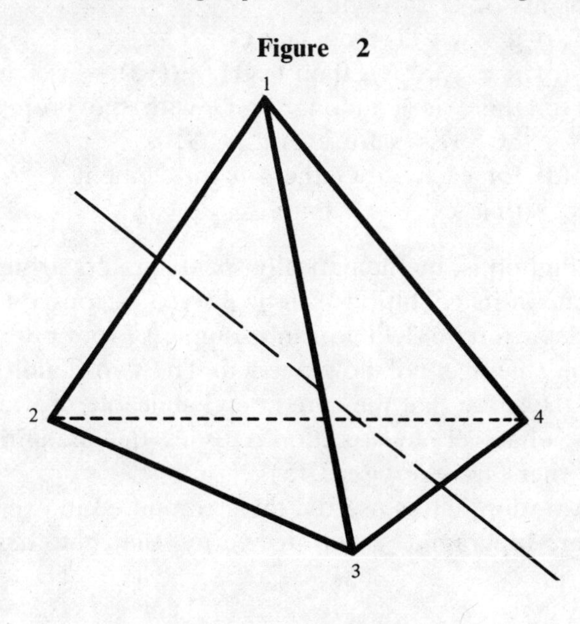

Section 2

Point Groups

By the time we have completed the table in exercise 1.4, we have in a sense moved beyond the six vertices (considered as points in space) to where we are considering only the motions r and c and their distinct products. These motions form a group, of which I give here two equivalent definitions.

(2.1) DEFINITION. A *group* G consists of a set, also denoted G, together with an operation ("multiplication") on pairs of elements of G satisfying:

(a) if $x, y \in G$, then $xy \in G$;
(b) if $x, y, z \in G$, then $(xy)z = x(yz)$;
(c) if $x, y \in G$, then there is a unique $u \in G$ with $xu = y$;
(d) if $x, y \in G$, then there is a unique $v \in G$ with $vx = y$.

(2.2) DEFINITION. A *group* G consists of a set, also denoted G, together with an operation ("multiplication") on pairs of elements of G satisfying:

(a) if $x, y \in G$, then $xy \in G$;
(b) if $x, y, z \in G$, then $(xy)z = x(yz)$;
(c) there is a unique $e \in G$ with the property that $ex = xe = x$ for every $x \in G$;
(d) for each $x \in G$ there is an element $x^{-1} \in G$ such that $xx^{-1} = x^{-1}x = e$.

The latter definition is, mathematically speaking, too strong, in the sense that some of its conditions can be proved as consequences of the others. (See exercise 2.11 for an example.) From our point of view such refinement is not now needed. The two definitions are *equivalent* in the sense that they are interchangeable; we can prove logically that whatever construction satisfies one definition also satisfies the other (see exercise 2.15).

As an illustration, let us use the table computed in exercise 1.4 for the square-bipyramid group to verify that the *associative property*, part (b) of each definition, holds in the specific instance $(rc)r = r(cr)$. We simply compute

$$(rc)r = (16)(25)(34) \cdot (2345) = (16)(35),$$
$$r(cr) = (2345) \cdot (16)(23)(45) = (16)(35).$$

Because of the associative property we can write a product like xyz unambiguously; it may be read as $(xy)z$ or $x(yz)$ at will.

Properties (c) and (d) of (2.1) say, in effect, that each row and each column of the table for the group will contain each element exactly once. (The verification of this fact has been left as an exercise.) However, writing down a table in which each row and each column contain each group element exactly once does *not* guarantee that the resulting structure will be a group. The associative property is tedious to verify in such a case.

Part (a) of either definition is called the *closure property*; the element e in (2.2c) is called the *identity* of the group; the element x^{-1} in (2.2d) is called the *inverse* of x.

An important observation about the inverse (see exercise 2.7) is:

(2.3) PROPOSITION. If G is a group and $x \in G$, then

$$(x^{-1})^{-1} = x.$$

The number of elements in a group, if finite, is called the *order* of the group. The group of transformations of the square bipyramid has order 8, and that of the trigonal bipyramid (introduced in exercise 1.5) has order 6. An example of an infinite group is the nonzero rational numbers together with ordinary multiplication.

Before proceeding to the main topic of this section, let us illustrate the definitions by examining some other easy examples of groups.

A group we shall encounter frequently is the *Klein 4-group*, denoted V_4, which has the four elements e, a, b, ab, and the "multiplication" table

	e	a	b	ab
e	e	a	b	ab
a	a	e	ab	b
b	b	ab	e	a
ab	ab	b	a	e

An important family of groups is the *finite cyclic* groups. The simplest example, denoted Z_2, consists of the two elements e, g, and is determined by the product $g^2 = e$, together with the requirements of (2.2c) for the identity. The order of Z_2 is, of course, 2. The group Z_3 consists of the elements e, g, g^2 with the multiplication given by

	e	g	g^2
e	e	g	g^2
g	g	g^2	e
g^2	g^2	e	g

If we use the elements e, r, r^2, r^3, we see that the rotations of the square bipyramid about the line Z of figure 1 form a cyclic group of order 4.

More generally, we may form the cyclic group Z_k of order k by taking the elements $e, g, g^2, \ldots, g^{k-1}$ and defining the multiplication by

$$g^i \cdot g^j = g^{i+j},$$

but in doing so we need to specify that $e = g^0 = g^k$, so that if the exponent $i+j$ is *greater than* $k-1$, we can write $i+j = k+m$ (that is, we take $m = i+j-k$), whence

$$g^{i+j} = g^{k+m} = g^k g^m = eg^m = g^m$$

with $0 \leq m < k$ as required. That Z_k is indeed a group depends upon properties of the integers that are tacitly assumed in this book; the interested reader may consult, for example, Dean [**5**], p. 43.

Now to connect the abstract definitions to the example of the square-bipyramid group, we may suppose that a group G and a set of points Ω are given (for example, the rigid symmetries of a plane or solid figure will always form a group). We then say that G *acts on* Ω if for each point $\alpha \in \Omega$ and for each $g \in G$, a correspondence is set up from α to a point denoted α^g such that

$$(2.4) \qquad \begin{aligned} (\alpha^g)^h &= \alpha^{(gh)} && \text{for } g, h \in G; \\ \alpha^e &= \alpha && \text{for the identity } e \text{ of } G. \end{aligned}$$

These two innocent-looking conditions may be quite difficult to fulfill if one simply chooses a group and a point set, but action of a group on a set usually arises in some natural way so that the conditions in (2.4) and the existence of G itself are automatically satisfied.

It will be helpful to consider the places in a point set Ω to which a given point α is eligible to go when we have a group G acting on Ω. For example, in the square-bipyramid group, the two vertices 1 and 6 can be interchanged (for example, by the motion c). The vertices 2, 3, 4, 5 can be rearranged among themselves in certain ways, though not so that 2 and 3 are at opposite ends of a diagonal of the square. Moreover, no motion interchanges any point of $\{1, 6\}$ with any point of the set $\{2, 3, 4, 5\}$. We refer to these sets of points as the orbits of α under G; more formally:

(2.5) DEFINITION. If the group G acts on the set Ω and if $\alpha \in \Omega$, then the *orbit* of α is $\{\alpha^g : g \in G\}$, denoted α^G. (The set-theoretic notation here means "the set of all points of the form α^g for each possible choice of g in G.")

A companion concept to the orbit arises when we observe that certain elements of G leave a given vertex fixed, as in the square bipyramid when $r^2c = (16)(24)$ moves neither 3 nor 5. We formulate this idea as:

(2.6) DEFINITION. If G acts on Ω and $\alpha \in \Omega$, then the *stabilizer of α in G* is $\{g \in G : \alpha^g = \alpha\}$, denoted G_α. Thus in the square-bipyramid group, $G_1 = \{e, r, r^2, r^3\}$ and $G_3 = \{e, r^2c\}$.

The same example leads us to the final consideration of this section. Recall that $rc \neq cr$. If, however, a group G does have the property that $xy = yx$ for *every* x and y in G, we say that G is *abelian* or *commutative*. The finite cyclic groups and the group V_4 are all abelian.

EXERCISES

2.1. Use the table of exercise 1.4 to verify that the associative property holds in the following specific instances:

$$\text{(a)} \qquad r(rc) = (rr)c$$

$$\text{(b)} \quad (r^2c)(rc) = r^2(c(rc))$$

(*Note:* In part (a), $(rr)c = r^2c$. In part (b), to find $r^2(c(rc))$, we must first find $c(rc)$ and then multiply on the left by r^2.)

2.2 Verify that the symmetries of the square bipyramid satisfy properties (a), (c), and (d) of (2.1) and (c) and (d) of (2.2).

2.3. Let 2, 4, 5 be the vertices so labeled in figure 1. Find G_2, G_4, and G_5.

2.4. Use (2.1) to show that in the multiplication table for a group, each element of the group appears exactly once in each row and exactly once in each column.

• *2.5.* Let G be a group, $x \in G$, and x^{-1} and x^* both be inverses of x. Prove that $x^{-1} = x^*$. Thus inverses are unique.

• *2.6.* Let G be a group and $g \in G$. Prove that $(g^{-1})^2 = (g^2)^{-1}$. (Because of this result we write simply g^{-2}. Similarly, $(g^{-1})^k = (g^k)^{-1}$ for any positive integer k, so we write g^{-k} without ambiguity.)

2.7. If G is a group and $x \in G$, prove that $(x^{-1})^{-1} = x$.

• *2.8.* Prove that the *cancellation laws* hold in a group G, that is, for a, b, $c \in G$:

$$\text{if} \quad ab = cb, \quad \text{then} \quad a = c;$$

$$\text{if} \quad ab = ac, \quad \text{then} \quad b = c.$$

2.9. In the notation of section 1, show that a cycle like (12) is its own inverse, that the inverse of (123) is (132), and that the inverse of (1234) is (1432). Generalize this result to a cycle with n distinct points appearing in it (each point appearing only once).

2.10. Countinuing exercise 2.9, find the inverse of (12)(34), of (12)(34)(56), and of (12)(345).

2.11. Show that the assumption of uniqueness in (2.2c) is not necessary; that is, show that if e and e^* are elements of a group G with the property that

$$ex = xe = x \quad \text{and} \quad e^*x = xe^* = x \quad \text{for every} \quad x \in G,$$

then $e = e^*$.

2.12. Find orbits and stabilizers for the vertices of the trigonal bipyramid introduced in exercise 1.5.

• *2.13.* In the Klein 4-group V_4, take a to be the rotation r^2 and b to be the motion c of the square-bipyramid group; verify that the set $\{e, r^2, c, r^2c\}$ has the table shown for V_4.

2.14. Let Ω be the power set of a group G (that is, the set of all subsets of G), and for $g \in G$, define

$$U^g = \{ug : u \in U\} \quad \text{for} \quad U \in \Omega.$$

Show that U^g defines an action of G on Ω. (This result is used in the appendix.)

2.15. Prove that each of (2.1) and (2.2) implies the other.

Section 3
Subgroups

If a group G acts on a set Ω, the stabilizer G_α of any α in Ω has an important property: it is a group in its own right. Whenever a subset H of a group G is also a group (using the same operation as in G), we call H a *subgroup.* For convenience, we introduce the following notation:

$$H \subseteq G \qquad H \text{ is a subset of } G,$$

$$H \leq G \qquad H \text{ is a subgroup of } G.$$

When we wish to specify that H is proper, that is, that $H \neq G$, we write $H \subset G$ or $H < G$, respectively.

Examples of subgroups are easy to find in the group table completed in exercise 1.4. Four such examples are $\{e\}$, $\{e, r, r^2, r^3\}$, $\{e, r^2\}$, and $\{e, c\}$. In fact, the group of the square bipyramid has a total of nine proper subgroups.

For the cyclic groups introduced in section 2 the situation varies with the order k of Z_k. The group Z_4 has proper subgroups $\{e\}$ and $\{e, g^2\}$, whereas the only proper subgroup of Z_5 is $\{e\}$.

Since the associative property refers only to the group operation itself, we need not check that it holds in a subset H when we try to prove that H is a subgroup. In fact, we have:

(3.1) PROPOSITION. Let H be a nonempty subset of G. Then H is a subgroup of G if and only if xy and x^{-1} are in H whenever x and y are in H.

PROOF. If $H \leq G$, the condition follows by (2.2). Conversely, in view of the remark above about associativity, we need show only that the identity e of G is in H. But if $x \in H$, then $x^{-1} \in H$, and $e = xx^{-1} \in H$ by the given condition.

A similar result is sometimes easier to use:

(3.2) PROPOSITION. Let H be a nonempty subset of G. Then H is a subgroup of G if and only if $xy^{-1} \in H$ whenever x and y are in H.

PROOF. Again, if $H \leq G$, the condition follows immediately. For the converse, we shall show that the condition in (3.1) follows from the given condition

$$(\star) \quad xy^{-1} \in H \text{ whenever } x, y \in H.$$

First, applying ($\star$) to x and x, we have $e = xx^{-1} \in H$. Now using e and x in ($\star$), we have $x^{-1} = ex^{-1} \in H$, as needed to cite (3.1). Similarly, $y^{-1} \in H$. Now by applying ($\star$) to x and y^{-1}, we have $xy = x(y^{-1})^{-1} \in H$, using (2.3). Hence by (3.1), $H \leq G$.

Now we are ready to prove:

(3.3) PROPOSITION. If G is a group acting on a point set Ω and if $\alpha \in \Omega$, then the stabilizer G_α is a subgroup of G.

PROOF. We observe that for any $x, y \in G_\alpha$, (2.4) assures that

$$\alpha = \alpha^e = \alpha^{(xx^{-1})} = (\alpha^x)^{x^{-1}} = \alpha^{x^{-1}}$$

and

$$\alpha^{(xy)} = (\alpha^x)^y = \alpha^y = \alpha,$$

so $G_\alpha \leq G$ by (3.1).

A useful observation is

(3.4) PROPOSITION. If x and y are in a group G, then

$$(xy)^{-1} = y^{-1}x^{-1}.$$

This result may be easily checked, given the fact that the uniqueness of the inverse in (2.2d) may be proved from other parts of the definition. (See exercise 2.5.)

We next form what is referred to in some quarters as "the set of formal products" of elements of H and K. Specifically, for subgroups H and K of a group G, the set HK consists of all products of the form hk, where the first term h is in H and the second k is in K. By the closure property, the set HK will always be a subset of G containing the identity e (because e is in H and in K and we may write $e = ee$). Of course a similar set KH can be formed. The significance may be seen in the following:

(3.5) PROPOSITION. Let H and K be subgroups of G; then HK is a subgroup of G if and only if $HK = KH$.

PROOF. First, assume that $HK = KH$. The crucial point here is that if $kh \in KH$, then kh is also in HK, and hence kh can be written as a product of an element h^* from H and an element k^* of K, that is, $kh = h^*k^*$ for one or more possible choices of h^* and k^*. (An illustration is given in exercise 3.14.) Now let h, $h' \in H$ and k, $k' \in K$. By applying the argument just given to k and h', we have $h^* \in H$ and $k^* \in K$ such that $kh' = h^*k^*$. Thus

$$(hk)(h'k') = h(kh')k' = h(h^*k^*)k' = (hh^*)(k^*k') \in HK.$$

In order to invoke (3.1), we need also to show that $(hk)^{-1} \in HK$. By applying the argument with which we began to k^{-1} and h^{-1}, we have $h^{**} \in H$ and $k^{**} \in K$ such that $k^{-1}h^{-1} = h^{**}k^{**}$; hence (using (3.4)), $(hk)^{-1} = k^{-1}h^{-1} = h^{**}k^{**} \in HK$. Thus by (3.1) HK is a subgroup of G. Conversely, if $HK \leq G$, then every element of HK is the inverse of a unique element of HK; hence for HK a list of inverses is also a complete list of elements, that is,

$$\begin{aligned} HK &= \{(hk)^{-1} : h \in H,\ k \in K\} \\ &= \{k^{-1}h^{-1} : h \in H,\ k \in K\} \\ &= KH \end{aligned}$$

since each element of H is an inverse of some element of H, and similarly for K. This completes the proof.

If $H \leq G$, we denote by Hg the set of all elements of the form hg, where $h \in H$ and g is some fixed element of G. Clearly, for each $h \in H$ we get such an hg. Moreover, if h, $h' \in H$ and $hg = h'g$, then $hgg^{-1} = h'gg^{-1}$, whence $h = h'$; thus each distinct element of H gives a distinct element of Hg. If H is finite, we conclude that H and Hg have the same number of elements. We call Hg a *coset of* H *in* G.

For any finite set S we shall denote by $|S|$ the number of elements in S. If S is infinite, we shall write $|S| = \infty$. For a finite group G, we have already introduced the term *order of* G for $|G|$. (The reader familiar with the concept of cardinality may interpret $|G|$ for infinite G as the *cardinal of* G in the results that follow.) From our discussion of cosets we may conclude:

(3.6) REMARK. If $H \leq G$ and $g \in G$, then $|H| = |Hg|$.

It is important to know that two cosets arising from the same subgroup will either coincide or be disjoint. Hence (using $\varnothing$ to denote the empty set) we prove:

(3.7) PROPOSITION. If $H \leq G$ and $x, y \in G$, then either $Hx = Hy$ or $Hx \cap Hy = \varnothing$.

PROOF. Suppose that $Hx \cap Hy \neq \varnothing$; then there exist h, $h' \in H$ such that $hx = h'y$ (that is, we have an element in the intersection). But

$hx = h'y$ gives $h^{-1}(hx)y^{-1} = h^{-1}(h'y)y^{-1}$, which simplifies to $xy^{-1} = h^{-1}h'$, and since the right-hand side is in H, we conclude that $xy^{-1} \in H$. Now for any element h^*x in Hx we have

$$h^*x = (h^*x)(y^{-1}y) = (h^*xy^{-1})y \in Hy$$

since $h^*(xy^{-1}) \in H$ by closure, and so $Hx \subseteq Hy$. If $h^*y \in Hy$, then

$$h^*y = (h^*y)(x^{-1}x) = (h^*(xy^{-1})^{-1})x \in Hx,$$

so $Hy \subseteq Hx$. But then the two cosets are equal.

The question of when two cosets do coincide is nicely characterized by

(3.8) PROPOSITION. If $H \leq G$ and $x, y \in G$, then $Hx = Hy$ if and only if $xy^{-1} \in H$.

PROOF. If $Hx = Hy$, then since $e \in H$, there exists $h \in H$ such that $ex = hy$ (because equality of sets means that the two have exactly the same elements). But then $xy^{-1} = h \in H$. The converse is included in the proof of (3.7).

Because of (3.6) and (3.7) we see that if $H \leq G$, then the cosets of H fill G. If G is finite, then G may be viewed as being subdivided into some number of cosets, each one having exactly as many elements as H, as suggested by figure 3. Thus we have proved:

(3.9) THEOREM OF LAGRANGE. If G is finite and $H \leq G$, then $|H|$ divides $|G|$.

Figure 3

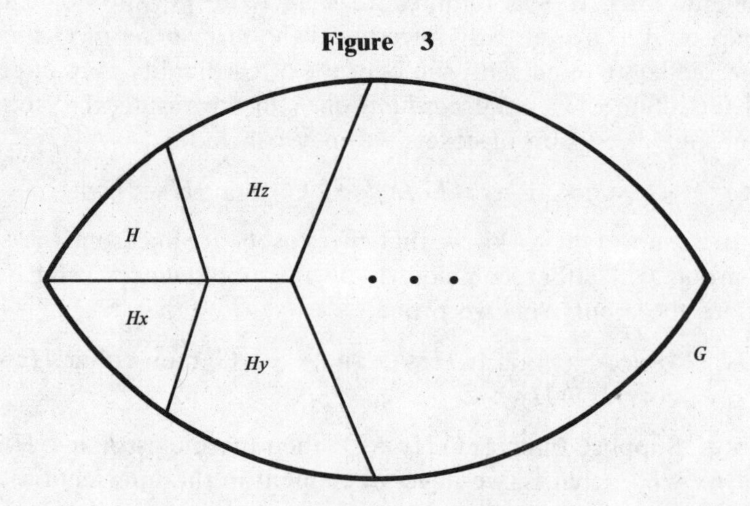

If $H \leq G$, then we denote by $[G:H]$ the number of cosets of H in G; this number is called the *index of H in G.* If the number of cosets is infinite, we write $[G:H]=\infty$. Note, however, that an infinite group G *may* have a subgroup H whose index is finite.

(3.10) PROPOSITION. If G is finite, then

$$|G|=|H|\cdot[G:H].$$

This result, which follows from (3.9), has the following interesting application to point groups:

(3.11) THEOREM. If G acts on Ω and $\alpha \in \Omega$, then

$$|G|=|G_\alpha|\cdot|\alpha^G|.$$

PROOF. From the preceding remarks, $|G|=|G_\alpha|\cdot[G:G_\alpha]$, so we need only to prove that $|\alpha^G|=[G:G_\alpha]$. Now for $x, y \in G$,

$$\begin{aligned} \alpha^x = \alpha^y \ &\text{iff}\ \alpha^{xy^{-1}} = \alpha && \text{by (2.4)} \\ &\text{iff}\ xy^{-1} \in G_\alpha \\ &\text{iff}\ G_\alpha x = G_\alpha y && \text{by (3.8);} \end{aligned}$$

that is, α^x and α^y are distinct exactly when the cosets $G_\alpha x$ and $G_\alpha y$ are distinct. Hence the number of points in α^G is $[G:G_\alpha]$.

This theorem may be used to compute the order of a point group. For example, consider the rigid symmetries of a cube, that is, the rearrangements of the vertices that can be performed without disassembling the cube. Consider the vertex numbered 1 in figure 4. Now $|1^G|=8$ since this vertex may occupy any one of the eight vertex locations. To find $|G_1|$, suppose a motion of G leaves vertex 1 fixed, and consider vertex 2. Since the cube is not to be taken apart, such a motion must carry vertex 2 to one of the three positions at opposite ends of the three edges that meet at 1, hence to vertex 2, 4, or 5. Thus $|G_1|=3$, since once we know where edge 1–2 lies, we have determined the position of the cube. By (3.11) we conclude that $|G|=8\cdot 3=24$, which we have established more easily than we could have by counting all possibilities. Once we have computed the order of G, the problem of finding all permutations of the vertices of the cube is simplified—we know how many transformations to look for. This example is pursued in exercises 3.4, 3.5, 3.6, and 3.11, and will be used again in section 9. (The reader interested in chemical applications may note that one

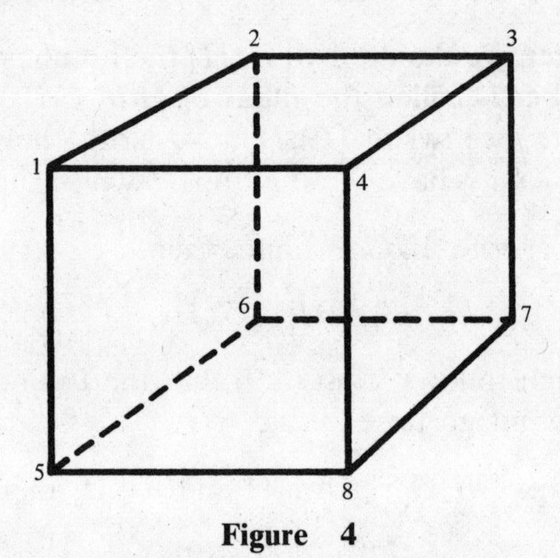

Figure 4

example in which the symmetries of the cube are useful is crystalline cesium chloride, CsCl, in which each cesium ion is located at the center of a cube of chloride ions.)

For applications in physical chemistry one usually allows, in addition to the rigid motions considered above, a reflection through a plane passing through the center of the cube and parallel to two faces, say to faces 1-2-3-4 and 5-6-7-8. In this case one easily checks that $|G_1| = 6$ and so $|G| = 48$, each new element in G_1 being obtained as the product of the new reflection, an element of the rigid symmetry group fixing some vertex other than 1 and 7, and the new reflection. This larger group of the cube is explored in exercises 3.7 through 3.9.

EXERCISES

- • *3.1.* Prove that if H and K are subgroups of G, so is $H \cap K$.
- *3.2.* Show by an example from the square-bipyramid group that $H \cup K$ need not be a subgroup of G even when H and K are subgroups.
- • *3.3.* Prove that if $K \leq H$ and $H \leq G$, then $K \leq G$.
- *3.4.* In the group G of rigid symmetries of the cube, let $x = (245)(386)$ and $r = (1234)(5678)$. Describe x and r geometrically as was done for the motions of the square-bipyramid group in section 1. Verify that $xr = r^3x^2$. (This verification *can* be done by direct computation, which is the hard way; an easier way is to find xr in terms of cycles, observe that xr is its own inverse, and apply (3.4).)
- *3.5.* Continuing exercise 3.4, let H be the subset of G consisting of all powers and all products of powers of x and r. By (3.1), H is a subgroup of G. Find 1^H, and then use (3.11) to prove that $H = G$.

3.6. Continuing exercises 3.4 and 3.5, observe that since $r^4 = x^3 = e$, there are twelve expressions of the form $r^i x^j$, with $i = 0, 1, 2, 3$ and $j = 0, 1, 2$. Prove first that any two such expressions are distinct, that is, that if $r^i x^j = r^m x^n$, then $i = m$ and $j = n$. Since $|G| = 24$, not every element of G can be written in this form. Prove that for *no* i and j can we have $x^2 r = r^i x^j$.

• *3.7.* Let $c = (15)(26)(37)(48)$ be the reflection mentioned in the text relative to figure 4. Then G^*, the group of the forty-eight symmetries of the cube, may be taken as the union of the group G of exercises 3.4–3.6 with the coset Gc (note that $G \leq G^*$). Find G_1^*.

• *3.8.* When a group K acts on a point set Ω, the *stabilizer of a subset* Δ of Ω is the subgroup whose members carry points of Δ only to points of Δ; it is denoted by K_Δ. For the group G^* of exercise 3.7, clearly $G_1^* \subseteq G_{\{1,7\}}^*$. However, the containment is in fact proper. Find $G_{\{1,7\}}^*$.

• *3.9.* In the notation of exercise 3.8, find the stabilizer of the set $\{1, 2\}$.

• *3.10.* Let G be a group and $g \in G$. Prove that the set of all powers of g and g^{-1}, which set we denote by $\langle g \rangle$, is a subgroup of G. The *order* of an element g is defined to be the order of $\langle g \rangle$, which may be finite or infinite. If G is a finite group, show that the order of any $g \in G$ divides $|G|$.

3.11. In the group G of exercise 3.4, find the orders of the elements x, r, and xr.

• *3.12.* In the notation of section 1, what is the order of the 2-cycle (12)? of the 3-cycle (123)? of the products (12)(34) and (123)(456)? (Recall exercises 2.9 and 2.10.) Tell why you would expect the order of the product (12)(345) to be 6, and verify that it is.

• *3.13.* If $H \leq G$ and $h \in H$, prove that $H = Hh$. Hence if $g \in G$, then $Hg = Hhg$.

3.14. In the notation of exercises 1.4 and 3.10, let $H = \langle r \rangle$ and $K = \langle c \rangle$. The table for the square-bipyramid group G gives G as HK (specifically, as ee, re, r^2e, r^3e, ec, rc, r^2c, r^3c); find an expression for each element of G as a member of KH. (*Note*: Exercise 3.6 shows that for the group of the cube, $\langle r \rangle \langle x \rangle \neq \langle x \rangle \langle r \rangle$; hence $\langle r \rangle \langle x \rangle$ is not a subgroup.)

Section 4

Homomorphisms and Normal Subgroups

Functions analogous to the linear transformations in linear algebra are basic to group theory. We call a function ϕ from a group G to a group G^* a *homomorphism* (in some books, *morphism*) if ϕ preserves products, that is, if

$$\phi(xy)=\phi(x)\phi(y)$$

for every $x, y \in G$. Clearly, if ϕ preserves products, it also preserves powers in the sense that $\phi(x^k)=(\phi(x))^k$ for positive k, and by exercise 2.6, for negative k as well. The notation

$$\phi: G \to G^*$$

will be used to mean that ϕ is a function from G to G^*. We prove first:

(4.1) Proposition. Let G and G^* be groups with identities e and e^*, respectively. If $\phi: G \to G^*$ is a homomorphism, then $\phi(e)=e^*$ and for each $x \in G$, $\phi(x^{-1})=\phi(x)^{-1}$.

Proof. Let $x \in G$; then $\phi(x)=\phi(ex)=\phi(e)\phi(x)$, and multiplying both sides by $\phi(x)^{-1}$ (which inverse exists because G^* is a group), we have $e^*=\phi(e)$. For the second part,

$$e^*=\phi(e)=\phi(xx^{-1})=\phi(x)\phi(x^{-1})$$

so $\phi(x^{-1})=\phi(x)^{-1}$ by uniqueness of inverses in G^* (see exercise 2.5).

If $\phi: G \to G^*$ is a homomorphism, we define the *kernel* of ϕ to be the set of all $x \in G$ such that $\phi(x)=e^*$, the identity of G^*. Note that the identity e of G is always in the kernel of ϕ, by (4.1). The fact that the kernel is a subgroup has been relegated to exercise 4.2.

The kernel has an additional property of basic importance for all of group theory. Let $\phi: G \to G^*$ be a homomorphism with kernel K and let $g \in G$. If $x \in K$, then

$$\phi(g^{-1}xg)=\phi(g^{-1})\phi(x)\phi(g)=\phi(g)^{-1}e^*\phi(g)=e^*,$$

so $g^{-1}xg \in K$. Now by $g^{-1}Kg$ we mean $\{g^{-1}xg : x \in K\}$, so what we have just shown is that

(4.2) $$g^{-1}Kg \subseteq K \quad \text{for all } g \in G.$$

In addition, with g^{-1} in place of g, we have

$$gKg^{-1} = (g^{-1})^{-1}Kg^{-1} \subseteq K.$$

Using this observation, we have

$$K = eKe = (g^{-1}g)K(g^{-1}g) = g^{-1}(gKg^{-1})g \subseteq g^{-1}Kg.$$

This containment now combines with (4.2) to prove that $K = g^{-1}Kg$ for every $g \in G$.

A subgroup N of a group G is called *normal* (or *invariant*) if $g^{-1}Ng = N$ for every $g \in G$. When N is a normal subgroup, we write $N \trianglelefteq G$; to indicate that N is both normal and proper in G, we may write $N \triangleleft G$. Obviously, for any group G, $\{e\} \trianglelefteq G$ and $G \trianglelefteq G$, and from the preceding paragraph together with exercise 4.2, we have:

(4.3) PROPOSITION. If $\phi: G \to G^*$ is a homomorphism, then the kernel of ϕ is a normal subgroup of G.

The *center* of a group G, denoted $C(G)$, consists of those elements of G each of which commutes with *every* element of G, that is,

$$C(G) = \{g \in G : gx = xg \quad \text{for all } x \in G\}.$$

If $g \in C(G)$ and $x \in G$, then $x^{-1}gx = x^{-1}xg = g \in C(G)$, hence:

(4.4) PROPOSITION. For any group G, the center $C(G)$ is a normal subgroup of G.

For some groups, the only normal subgroups are $\{e\}$ and the group itself; such groups are called *simple*. The easiest example of a simple group is a cyclic group of order p, where p is some prime (positive) integer. Since the order of Z_p is the prime p, by (3.9) the only possible subgroups are those having order 1 (the identity alone) or p (the group itself).

EXERCISES

• *4.1.* If $\phi: G \to G^*$ is a homomorphism, then the image $\phi(G)$ is a subgroup of G^*.

4.2. Let $\phi: G \to G^*$ be a homomorphism. Show that the kernel of ϕ is a subgroup of G.

4.3. If G is abelian, then every subgroup of G is normal.

4.4. If $g \in C(G)$, what can you say about the row and the column corresponding to g in a multiplication table (arranged as in section 1)?

4.5. If G is abelian, what is $C(G)$?

4.6. Find the center of the square-bipyramid group.

4.7. Let G be any group, $H \leq G$, and $N \trianglelefteq G$. Prove that $HN \leq G$.

• *4.8.* Let G be any group, N and M be normal subgroups of G, and $N \cap M = \{e\}$. Prove that $nm = mn$ for every $n \in N$ and $m \in M$. Then explain why this result does *not* imply that G must be abelian.

• *4.9.* In section 3 we decomposed a group G into *right* cosets Hg of a subgroup H; we can also use *left* cosets gH. Show that if

$$G = Hg_1 \cup Hg_2 \cup \ldots \cup Hg_n,$$

then we also have

$$G = g_1^{-1}H \cup g_2^{-1}H \cup \ldots \cup g_n^{-1}H.$$

Next, prove that H is normal in G if and only if $gH = Hg$ for every $g \in G$. Finally, give an example to show that gH and Hg *may* be neither coinciding nor disjoint (compare (3.7)). Left cosets will be used in section 17.

Section 5

Isomorphisms

Normal subgroups also arise in the context of an important action of a group G on the set of its own elements.* For each $x, g \in G$, we set

(5.1) $$x^g = g^{-1}xg.$$

Now by (5.1) and (3.4),

$$(x^g)^h = (g^{-1}xg)^h = h^{-1}(g^{-1}xg)h = (gh)^{-1}x(gh) = x^{gh}$$

and

$$x^e = e^{-1}xe = exe = x;$$

hence the conditions (2.4) are satisfied, and (5.1) defines an action of the group G on the set G. We say that *G acts on itself by conjugation*, and (5.1) is called *conjugation* of the element x by g. Any element of the form $g^{-1}xg$ is therefore called a *conjugate of* x, and the orbit of x

$$x^G = \{g^{-1}xg : g \in G\}$$

is the *conjugate class of* x. The stabilizer

$$G_x = \{g \in G : g^{-1}xg = x\}$$

is called the *centralizer* of x in G.

Using (2.3), we can observe:

(5.2) PROPOSITION. If G is a group and $x, g \in G$, then conjugation by g^{-1} takes x to gxg^{-1}.

Next we need to generalize the concepts of orbit and stabilizer from a single element of G to a subset (compare exercises 3.8 and 3.9). Let S be a subset (possibly but not necessarily a subgroup) of G; we take the orbit of the set S under G to be

$$S^G = \{g^{-1}sg : s \in S, g \in G\}$$

* Recall that in defining a group at (2.1) and (2.2), we allowed the letter G to refer either to the set or to the set-with-operation; this ambiguity is conventional and should cause no confusion.

and the stabilizer of S to be

$$N_G S = \{g \in G : g^{-1}Sg = S\};$$

the latter is called the *normalizer* of S in G. Note that if $g \in N_G S$, then for each $s \in S$, the element $g^{-1}sg$ is in S, but $g^{-1}sg$ need not be equal to s itself, unless S contains only one element. For example, let $S = \{r, r^3\}$ in the group of the square bipyramid; then one can check that $c^{-1}rc = r^3$, and thus $c \in N_G S$, even though conjugation by c does not carry r to itself. Thus a companion concept is the *centralizer* of a subset S, denoted $Z_G S$ and defined by

$$Z_G S = \{g \in G : g^{-1}sg = s \quad \text{for all } s \in S\}.$$

If $S = G$, then its centralizer is the center of G, that is,

$$Z_G G = C(G).$$

Note also that if e is the identity of G, then $Z_G\{e\} = G$. Of course, if G is abelian, then for any $x, g \in G$,

$$x^g = g^{-1}xg = g^{-1}gx = x,$$

so $Z_G S = G$ for every subset S of an abelian group G.

It is easy to verify that conjugation is a homomorphism of G, for any group G. In addition, if $g^{-1}xg = g^{-1}yg$, then $x = y$ by exercise 2.8, whence conjugation is a $1{:}1$ function on G. Moreover, for any $y \in G$, gyg^{-1} is an element of G with the property that $g^{-1}(gyg^{-1})g = y$; hence conjugation is also an onto function. (The terms *injective* and *surjective* are often used for *1:1* and *onto*, respectively.)

If a homomorphism $\phi : G \to G^*$ is a $1{:}1$ function, then ϕ is called an *isomorphism*. Groups G and G^* are *isomorphic*, denoted

$$G \cong G^*,$$

if there is an isomorphism of G *onto* G^*. An *automorphism* is an isomorphism of a group *onto* itself. Thus conjugation by some element g is an example of an automorphism of a group G; this function is important enough that we give it the special name of *inner automorphism*. An automorphism that is not an inner automorphism is called (not surprisingly) *outer*. Since every inner automorphism of an abelian group is necessarily the identity function, all automorphisms of an abelian group are outer except for the identity function.

Exercise 2.13 includes an incidental example of an isomorphism. To be more specific, let

(5.3) $\phi(a)=r^2, \quad \phi(b)=c, \quad \phi(ab)=r^2c, \quad \phi(e)=e;$

then ϕ is clearly a 1:1 onto function from the Klein 4-group $V_4=\{e, a, b, ab\}$ of section 2 to the subgroup $\{e, r^2, c, r^2c\}$ of the square-bipyramid group. That ϕ is a homomorphism follows from the observation that the multiplication tables are identical except for the choice of letters, and thus all products are preserved. Fortunately, it is not necessary in general to complete multiplication tables for two groups in order to check that products are preserved, as we shall see.

Note that we have required an automorphism to be an onto function but that we have not made this stipulation for isomorphisms. Thus if $\phi: G \rightarrow G^*$ is an isomorphism, the image $\phi(G)$ may be a proper subset of G^*, but since (by exercise 4.1) $\phi(G)$ is a subgroup of G^*, we conclude that $G \cong \phi(G)$. Note that for the homomorphism ϕ defined by (5.3) the image $\phi(V_4)$ is a proper subgroup of the square-bipyramid group. We shall see in section 7 that if G is infinite, then G may even be isomorphic to a proper subgroup of itself.

The concept of conjugation bears an interesting connection to the idea of action of a group G on an *arbitrary* set S, as follows:

(5.4) THEOREM. Let G act on S, let $\alpha, \beta \in S$, and let $g \in G$. Then

$$\alpha^g = \beta \quad \text{implies that } g^{-1}G_\alpha g = G_\beta.$$

PROOF. Let $h \in G$. Then

$$\begin{aligned} h \in G_\beta \quad &\text{iff} \quad \beta^h = \beta \\ &\text{iff} \quad \alpha^{gh} = \alpha^g \qquad \text{using (2.4)} \\ &\text{iff} \quad \alpha^{ghg^{-1}} = \alpha \\ &\text{iff} \quad ghg^{-1} \in G_\alpha \\ &\text{iff} \quad h \in g^{-1}G_\alpha g. \end{aligned}$$

This chain of logical equivalences shows that $G_\beta = g^{-1}G_\alpha g$.

We may paraphrase (5.4) as follows:

(5.4′) THEOREM. Let the group G act on the point set S; if the group element g carries point α to point β, then the stabilizer of β is the conjugate by g of the stabilizer of α.

An example is given in exercise 5.13. From the point of view of applications, once we have the stabilizer of any point α in the set S, we may use (5.4) to find the stabilizer of any other point in the orbit of α; in particular, if G is finite, then since conjugation is a 1:1 function the stabilizers of all points in a given orbit will have the same order. (On this topic see exercise 5.12.)

There are three classical theorems relating to homomorphisms and isomorphisms. For the first of these, recall from section 3 that if $H \leq G$, then G is subdivided into a set of disjoint cosets of H, each having the form Hg for some $g \in G$. We denote this set by G/H and call it a *factor set.* Note that each element Hg of G/H is a *subset* of G, as suggested by figure 3 in section 3. We can now state:

(5.5) HOMOMORPHISM THEOREM I. Let $\phi: G \rightarrow G^*$ be a homomorphism of G *onto* G^* and K be the kernel of ϕ. Then $K \trianglelefteq G$, and G/K forms a group under the operation $(Kg)(Kh) = K(gh)$. Moreover, $G/K \cong G^*$.

PROOF. That $K \trianglelefteq G$ was proved in section 4. The proof that G/K with the operation indicated satisfies the group axioms is left as exercise 5.4. In this instance, however, we must check one further fact in order to assert that G/K is a group. Note that even when $x \neq y$ in G it may happen that $Kx = Ky$ if $xy^{-1} \in K$. Thus we are obliged to check that the operation specified by

$$(Kg)(Kh) = K(gh)$$

is *well defined* for G/K, that is,

(5.6) if $Kx = Ky$ and $Ku = Kv$, then $(Kx)(Ku) = (Ky)(Kv)$.

No confusion results if we write Kgh to mean $K(gh)$. By (3.8), the assertion (5.6) is equivalent to

(5.6′) if $xy^{-1}, uv^{-1} \in K$, then $(Kx)(Ku) = (Ky)(Kv)$.

Now if xy^{-1} and uv^{-1} are in K, then their inverses yx^{-1} and vu^{-1} are also in K, and since $K \trianglelefteq G$, $x(vu^{-1})x^{-1}$ is in K (see (5.2)). Now

$$\begin{aligned}(Kx)(Ku) &= Kxu \\ &= K(xvu^{-1}x^{-1})(xu) \quad \text{by exercise 3.13} \\ &= Kxv \\ &= K(yx^{-1})(xv) \qquad \text{again by exercise 3.13} \\ &= Kyv \\ &= (Ky)(Kv),\end{aligned}$$

which completes the check that the operation is well defined. Finally, to see that $G/K \cong G^*$, we define $\sigma: G/K \to G^*$ by $\sigma(Kg) = \phi(g)$ for each coset Kg. Then

$$\begin{aligned}\sigma((Kg)(Kh)) &= \sigma(Kgh)\\ &= \phi(gh)\\ &= \phi(g)\phi(h)\\ &= \sigma(Kg)\sigma(Kh),\end{aligned}$$

so σ is a homomorphism from G/K to G^*. Now σ is onto since $\phi(G) = G^*$. That σ is 1:1 and well defined follows from

$$\begin{aligned}\sigma(Kg) = \sigma(Kh) \quad &\text{iff} \quad \phi(g) = \phi(h)\\ &\text{iff} \quad \phi(g)\phi(h)^{-1} = e^*\\ &\text{iff} \quad \phi(gh^{-1}) = e^*\\ &\text{iff} \quad gh^{-1} \in K\\ &\text{iff} \quad Kg = Kh \qquad \text{by (3.8)},\end{aligned}$$

where e^* is the identity of G^*. This completes the proof of (5.5).

Since the factor set G/K is a group when $K \trianglelefteq G$ (under the operation specified in (5.5)), we call it a *factor group* or, to be more specific, the *factor group of G modulo K.*

For an example, we consider first the group of symmetries of the square bipyramid, which may be written as

$$\{e, r, r^2, r^3, c, rc, r^2c, r^3c\}.$$

This group is also called the *dihedral group* of order 8 and is denoted D_4. We also need $V_4 = \{e, a, b, ab\}$; no harm is done by using e to represent the identity of both groups. Let $\phi(r) = a$ and $\phi(c) = b$; then since ϕ is to preserve products, the rest of its values have already been determined, for example, $\phi(r^2) = a^2$ and $\phi(rc) = ab$. It is easy, if a bit tedious, to verify that ϕ is a homomorphism onto and that its kernel $K = \{e, r^2\}$. Then

$$\begin{aligned}Kr &= \{r, r^3\} = Kr^3,\\ Kc &= \{c, r^2c\} = Kr^2c,\\ Krc &= \{rc, r^3c\} = Kr^3c,\\ K &= Kr^2;\end{aligned}$$

hence we may write $D_4/K = \{K, Kr, Kc, Krc\}$. We have proved that $D_4/K \cong V_4$ and also written D_4/K in such a form as to make the isomorphism visually apparent.

Conversely to (5.5), let N be any normal subgroup of a group G, and define

$$\eta: G \to G/N \quad \text{by} \quad \eta(g) = Ng.$$

As in (5.5), take the operation in G/N to be

$$(Nx)(Ny) = Nxy \quad \text{for} \quad x, y \in G.$$

Then it is easy to check that η is a homomorphism onto with kernel N. The group G/N is again called the *factor group* or *quotient group* of G by N.

If H is a subgroup of G but *not* a normal subgroup, then the factor set G/H does *not* form a group under the operation of (5.5). An illustration is given in exercise 5.6.

In the example of $D_4/K \cong V_4$, the factor group has the table

	K	Kx	Ky	Kxy
K	K	Kx	Ky	Kxy
Kx	Kx	K	Kxy	Ky
Ky	Ky	Kxy	K	Kx
Kxy	Kxy	Ky	Kx	K

Two observations should be apparent:

(5.7) PROPOSITION. If $K \trianglelefteq G$, then K is the identity of G/K.

(5.8) PROPOSITION. Two groups are isomorphic if they are in fact the same (in elements and multiplication table) except for the symbols used to write them.

The second classical theorem on homomorphisms may be stated in various ways; we use the following formulation:

(5.9) HOMOMORPHISM THEOREM II. Let $\phi: G \to G^*$ be a homomorphism of G *onto* G^* with kernel K. Then for each $H \leq G$ such that $K \leq H$, we have $H/K \cong \phi(H)$. Conversely, if $H^* \leq G^*$, then there is a subgroup H of G containing K with $\phi(H) = H^*$ and $H/K \cong H^*$. Moreover, if $K \leq H \leq G$, then $H \trianglelefteq G$ if and only if $\phi(H) \trianglelefteq G^*$.

This theorem says in effect that the pattern of subgroups of G^* is precisely that of G if we ignore all subgroups of G not containing the kernel K of ϕ. In terms of the preceding example, we have the pattern of inclusions in figure 5; the solid lines show the pattern of inclusions that are carried over by ϕ, and the dotted lines indicate inclusions that are suppressed because they do not

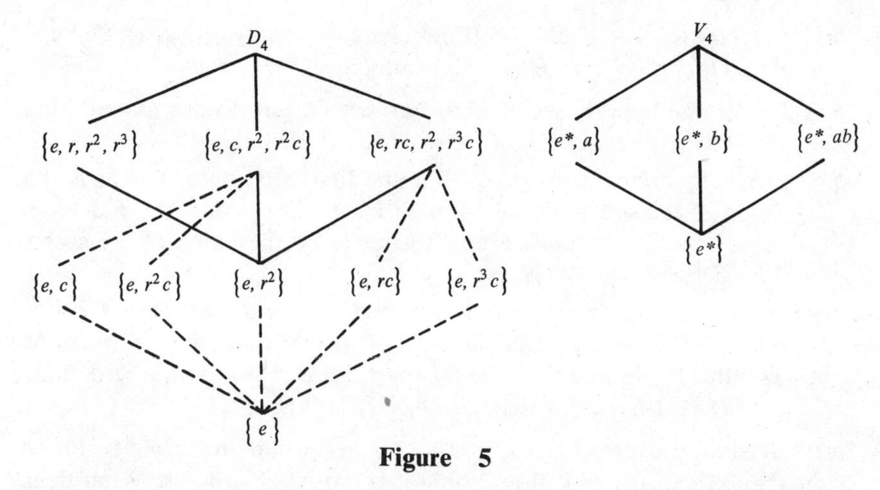

Figure 5

pertain to subgroups containing the kernel $\{e, r^2\}$. The proof of (5.9) is primarily of interest in studies in abstract group theory; for completeness, a sketch is given in exercise 5.8.

To conclude this section, let us consider a statement of the third classical theorem on homomorphisms:

(5.10) HOMOMORPHISM THEOREM III. Let G be a group, $H \leq G$, and $N \trianglelefteq G$. Then $HN \leq G$, $N \trianglelefteq HN$, $H \cap N \trianglelefteq H$, and

$$HN/N \cong H/(H \cap N).$$

This theorem is especially useful in the study of group extensions, in which, given the kernel N and an isomorphic copy of the factor group G/N, one attempts to find the groups G that can have the given kernel and factor group.

EXERCISES

- •*5.1.* If S is any subset of a group G, prove that $N_G S$ and $Z_G S$ are subgroups of G. If, in addition, $S \leq G$, prove that $S \trianglelefteq N_G S$. (*Note:* The latter result requires proof of both containment and normality.)
- *5.2.* If S and N are subgroups of G and $S \trianglelefteq N$, then $N \leq N_G S$.
- *5.3.* Show that a homomorphism $\phi: G \to G^*$ is an isomorphism if and only if the kernel of ϕ consists of the identity of G alone.
- *5.4.* Prove that G/K in (5.5) satisfies the axioms for a group.
- *5.5.* Verify that the conclusions of (5.5) hold in the example of D_4 and V_4.

5.6. Let $H=\{e, c\}$ and $G=D_4$. Find elements g_1, g_2, g_3, g_4 in D_4 such that $Hg_1=Hg_2$ and $Hg_3=Hg_4$ but $Hg_1g_3 \neq Hg_2g_4$.

5.7. Under the hypotheses of (5.9), let $L \leq G$, but do not assume that $K \leq L$. Prove that $KL \leq G$ and $\phi(KL) \cong KL/K$.

5.8. Prove theorem (5.9). (*Hint*: For the first part, note that H is the union of the set of cosets Kh with $h \in H$. Consider what ϕ does to these cosets. For the second part, let H be the union of the cosets Kh such that $\phi(h) \in H^*$.)

5.9. Prove theorem (5.10). (*Hint*: The first assertion is exercise 4.7. For the second, let $hn \in HN$ and $m \in N$; prove that $(hn)^{-1}m(hn) \in N$. The third is similar. For the last part, let $\phi: H \to HN/N$ be defined by $\phi(h)=Nh$ (recall that $HN=NH$) and use (5.5).)

• *5.10.* (Based on exercise 3.10.) Let G be any group and $x, y \in G$. Prove that x is of order k if and only if $y^{-1}xy$ is of order k. What does this result imply about the orders of the elements found in a given conjugate class? What can you say about the number of elements to be found in a given conjugate class? If $x \in C(G)$, what is the conjugate class of x?

• *5.11.* Let $\phi: G \to G^*$ be an isomorphism and x be an element of G having order k. Prove that $\phi(x)$ has order k. How does the answer change if ϕ is taken only to be a homomorphism?

• *5.12.* Let H be a subgroup of a group G and $g \in G$. Prove that $g^{-1}Hg \leq G$ and that $|H|=|g^{-1}Hg|$. When H and K are subgroups of G with the property that for some $g \in G$, $K=g^{-1}Hg$, H and K are called *conjugate subgroups.*

5.13. Illustrate (5.4′) for the group of the square bipyramid by taking α and β to be vertices 2 and 5, respectively. Specifically, find G_2 and G_5 (see exercise 2.3), find the two elements of G which carry 2 to 5, and show that conjugation by each of these elements transforms G_2 into G_5.

5.14. Let $G=\{e, g, g^2, \ldots, g^8\}$ be cyclic of order 9 and $G^*=\{e, h, h^2\}$. Prove that $\phi(g^a)=h^{2a}$ for $0 \leq a \leq 8$ defines a homomorphism of G onto G^* but that σ is not an isomorphism.

5.15. Prove that if ϕ is an automorphism of G, then $\phi(C(G))=C(G)$.

5.16. (For readers familiar with equivalence relations.) Let G be a group acting on a set S, and for $x, y \in S$ define $x \sim y$ if and only if there exists $g \in G$ such that $y=x^g$. Prove that $\sim$ is an equivalence relation and that the orbit of x is an equivalence class. What does this result say when the action is conjugation on the set G?

Section 6
Sylow Subgroups

The serious study of finite groups began in 1872 with the publication of some remarkable results by Ludvig Sylow. In this section we shall consider these theorems, some related results, and some applications. To concentrate on how the Sylow theorems are used, I omit the proofs from this section and refer the interested reader to the appendix.

Throughout this section, let G be a *finite* group, and for positive integers m and n, let $m \mid n$ denote "m divides n" and $m \nmid n$ denote "m does not divide n."

As is common in group theory, we now use the symbol 1 rather than e for the identity of G.

The Theorem of Lagrange (3.9) says that if $H \leq G$, then $|H|$ divides $|G|$. The logical converse of this result is: if k divides $|G|$, then G has a subgroup of order k; unfortunately, this statement is false, and in section 7 we shall find a group A_5 of order 60 that has no subgroup of order 15, 20, or 30. The first Sylow theorem is a partial converse to the Theorem of Lagrange, though not the strongest result in this direction.

(6.1) SYLOW THEOREM I. Let p be a prime, $|G| = p^e q$, where $p \nmid q$. Then G contains a subgroup of order p^e, which is called a *Sylow p-subgroup.*

Note that in (6.1) the order of a Sylow p-subgroup is p^e, not p itself (unless $e = 1$).

Note also that it is possible to have $e = 0$ in (6.1) in the event p is a prime that does not divide $|G|$; then of course $q = |G|$, and $\{1\}$ qualifies as a Sylow p-subgroup of G. In practice, we shall refer to a Sylow p-subgroup only when p is a prime actually dividing $|G|$, so that $e \geq 1$.

Thus the group A_5 mentioned above, whose order is $60 = 2^2 \cdot 3 \cdot 5$, must have subgroups of orders 4, 3, and 5. We shall see that it also has subgroups of all possible orders *except* 15, 20, and 30, namely of orders 2, 6, 10, 12, and (of course) 1 and 60.

Groups whose orders are powers of a single prime p have many interesting properties; we call such groups *p-groups* (hence the term *Sylow p-subgroup*). Among these properties are:

(6.2) PROPOSITION. The center of a finite p-group has order greater than 1.

(6.3) PROPOSITION. If G is a finite p-group and k divides $|G|$, then G has a subgroup of order k.

The proofs of these two propositions are given along with those of the Sylow theorems in the appendix; for the present we shall observe that as a consequence of (6.1) and (6.3) with exercise 3.3, we have:

(6.4) COROLLARY. If p is prime and $|G| = p^e q$ with $p \nmid q$, then G has subgroups of orders $1, p, p^2, \ldots, p^{e-1}, p^e$.

For infinite groups, the situation is quite different. Recall from exercise 3.10 that the *order of an element* g of a group G is the smallest nonnegative integer k such that $g^k = 1$. An infinite group is a p-group if each of its elements has a finite order that is a power of the given prime p (not necessarily the same power for all elements!). Proposition (6.2) is *false* for infinite p-groups.

The second of the Sylow theorems relates the concepts of p-subgroup and conjugacy. Note that if G satisfies the hypotheses of (6.1) and H is a subgroup of G with $|H| = p^k$, then $k \leq e$ since p^k must divide p^e, by (3.9).

(6.5) SYLOW THEOREM II. Let $|G| = p^e q$ with p prime and $p \nmid q$. Let $H \leq G$ with $|H| = p^k$, and let P be *any* Sylow p-subgroup of G. Then there exists an element $g \in G$ such that $H \leq g^{-1}Pg$.

Observe first that since conjugation is an automorphism of G and hence a 1:1 function, $g^{-1}Pg$ contains p^e elements. It is easy to show that a conjugate of a subgroup is again a subgroup (see exercise 5.12); hence $g^{-1}Pg$ is also a Sylow p-subgroup of G, and (6.5) says that a subgroup of order p^k is contained in *some* Sylow p-subgroup. More important, if $k = e$ in the hypotheses of theorem (6.5), then H and P are two arbitrary Sylow p-subgroups of G. Hence:

(6.6) COROLLARY. Any two Sylow p-subgroups of G (for the same prime p) are conjugate.

Thus conjugacy induces an action of the group G on the set of Sylow p-subgroups of G, and (6.6) guarantees that (for a given p) all of the Sylow p-subgroups lie in a single orbit with respect to this action. Then from the definition of normalizer in section 5 together with (3.10), we have:

(6.7) PROPOSITION. The number of distinct Sylow p-subgroups of G is equal to the index in G of the normalizer of any one Sylow p-subgroup. Moreover, the normalizers of two Sylow p-subgroups have the same order.

A more specific result than (6.7) on the number of Sylow p-subgroups is:

(6.8) SYLOW THEOREM III. Let $|G| = p^e q$ with p prime and $p \nmid q$. Then the number s_p of distinct Sylow p-subgroups of G is a divisor of q, and p divides $(s_p - 1)$.

In the language of elementary number theory, the conclusion of (6.8) may be written as

$$s_p \mid q, \quad \text{and} \quad s_p \equiv 1 \pmod{p}.$$

These theorems give a remarkable amount of information about the structure of a finite group of a given order. For a rather surprising example, let $|G| = 35 = 5 \cdot 7$. First let $p = 5$; then $e = 1$ and $q = 7$. Then s_5 divides 7 and so must be 1 or 7, but also 5 divides $(s_5 - 1)$, which means that 5 must divide 0 or 6. Hence $s_5 = 1$. A similar argument leads to $s_7 = 1$. Now since any conjugate of a Sylow p-subgroup is also a Sylow p-subgroup, from (6.6) we have:

(6.9) PROPOSITION. The group G has a normal Sylow p-subgroup if and only if $s_p = 1$.

Thus in our example of $|G| = 35$, G must have normal subgroups H_5 and H_7 of order 5 and 7, respectively. Now by exercise 3.10, every element of H_5 except the identity has order 5 (since only the identity can have order 1), and every nonidentity element of H_7 has order 7. Therefore $H_5 \cap H_7 = \{1\}$. Next we need the result of exercise 4.8, restated here as:

(6.10) PROPOSITION. If $N, M \trianglelefteq G$ and $N \cap M = \{1\}$, then $nm = mn$ for every $n \in N$ and $m \in M$.

This result holds for infinite as well as finite groups; note, however, that it says nothing whatever about whether the subgroups N and M themselves are abelian, but only that each element of N commutes with every element of M. In particular, if $n \in N$ and $m \in M$, then

$$(6.11) \qquad (nm)^2 = n(mn)m = n(nm)m = n^2m^2,$$

and similarly for higher powers of nm.

Now to apply these results to our group G of order 35, let $x \neq 1$ in H_5 and $y \neq 1$ in H_7. Then

$$H_5 = \{1, x, \ldots, x^4\} \quad \text{and} \quad H_7 = \{1, y, \ldots, y^6\}.$$

By (6.11), $(xy)^k = x^k y^k$, so $(xy)^k = 1$ if and only if both 5 and 7 divide k. But then 35 divides k, so the powers of xy are all of the elements of G, and G is cyclic. We conclude that, except for isomorphic equivalents, there is *only one* group of order 35.

EXERCISES

- • *6.1.* Apply the techniques at the end of the section to show that there is only one group of order 15. Why can the same techniques *not* be used to argue that a group of order 21 must be cyclic? (*Note*: In fact there *are* noncyclic groups of order 21.)
- *6.2.* Determine whether a group of order 39 can be simple.
- • *6.3.* Our discussion of groups of order 35 uses the fact that any two finite cyclic groups of the same order are isomorphic. Prove this fact.
- *6.4.* Let G be any group, $N, M \leq G$, and suppose that $|N|$ and $|M|$ are relatively prime. Then $N \cap M = \{1\}$, but show by example that $nm = mn$ need *not* hold for every $n \in N$ and $m \in M$.
- *6.5.* Let G be an abelian group and p a prime dividing $|G|$; prove that G has only one Sylow p-subgroup.
- *6.6.* Find the order of each element of the cyclic group Z_{12}.
- *6.7.* Continuing exercise 6.6, find the Sylow 2- and 3-subgroups of Z_{12}.
- *6.8.* For the group of rigid symmetries of the tetrahedron (see exercise 1.7) find the unique Sylow 2-subgroup and the four Sylow 3-subgroups.
- *6.9.* Prove that a group of order 12 cannot be simple. (*Hint*: Show that if $s_3 = 4$, then G must have eight elements of order 3 and that this fact implies $s_2 = 1$.)
- *6.10.* Let P be a Sylow p-subgroup of G, $N = N_G P$, and H a subgroup such that $N \leq H \leq G$. Prove that $N_G H = H$.

Section 7
Further Examples

Before turning to representation theory, let us consider some additional examples of specific groups and the useful technique of presenting a group in terms of generators and relations.

The set of indicated powers (positive, negative, and zero) of a single symbol x forms an *infinite cyclic group* under the operation of multiplication, using the usual laws of exponents

$$x^i \cdot x^j = x^{i+j},$$

$$x^0 = 1 \quad \text{(the identity).}$$

Note that, as in exercise 2.6, $(x^i)^{-1} = x^{-i}$, for any integer i. In contrast to the reduction $g^k = 1$ in the *finite* cyclic group Z_k, our infinite cyclic group has $x^k = 1$ *only* when $k = 0$. Moreover, if $x^k = x^m$, then $x^{k-m} = 1$, whence $k = m$.

Now suppose that we also form an infinite cyclic group from the powers of an element y that is distinct from all powers of x. Then $\phi(x^k) = y^k$ defines an isomorphism of the first infinite cyclic group onto the second, and we conclude that:

(7.1) PROPOSITION. Any two infinite cyclic groups are isomorphic.

The reader may check that $\psi(x^k) = y^{-k}$ also defines an isomorphism onto.

In view of (5.8) and (7.1), we are justified in speaking of *the* infinite cyclic group; we denote it by Z_∞. By exercise 6.3, we may also speak of *the* cyclic group of order k for any positive integer k. (Although we did not formally define Z_k for $k = 1$, our definition of order of an element allows us to take $Z_1 = \{1\}$.) We also observe that if K is the subgroup of Z_∞ consisting of all powers of x^k for fixed k, then $Z_\infty/K \cong Z_k$.

Whenever two groups are isomorphic, we say that they represent the same *abstract group*, recalling (5.8). Thus we say that there is only one abstract group that is infinite and cyclic, and only one cyclic group of each finite order k. Similarly, from section 6, there is only one abstract group of order 35.

The group Z_∞ has an unusual property related to the Dedekind characterization of an infinite set, which is that an infinite set is one that can be put into 1:1 correspondence with one of its proper subsets. Let

$$\phi(x^i) = x^{2i}$$

for each integer i; then ϕ is an isomorphism of Z_∞ onto a proper subgroup of itself, namely, the subgroup consisting of all even powers of x. Hence Z_∞ is isomorphic to a *proper* subgroup of itself.

We have considered in previous sections and their exercises the rigid symmetries of various solid figures. We shall now derive an important family of groups from plane figures. For any positive integer $n > 2$, consider the regular polygon of n sides (equilateral triangle, square, regular pentagon, etc.). We can form a group of symmetries of the vertices of the polygon in the plane. For the square, the group G of symmetries will have order 8 by (3.11), since the orbit of any vertex is the set of all 4 positions, and the order of its stabilizer is clearly 2. (For example, in figure 6, vertex 2 must either remain fixed under an element of the stabilizer G_1 or must go to position 4.) Let

$$x = (1234) \quad \text{and} \quad y = (14)(23).$$

Then one can easily verify that $x^4 = e$ (the identity), $y^2 = e$, and $yx = x^3y$. If we consider the transformations denoted by

$$e,\ x,\ x^2,\ x^3,\ y,\ xy,\ x^2y,\ x^3y,$$

Figure 6

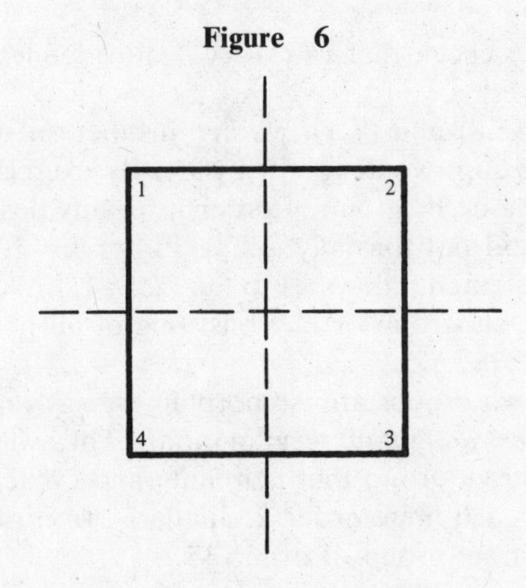

we find that they are precisely the eight distinct elements of G. Moreover, if we define

$$\phi(x^i y^j) = r^i c^j \quad \text{for} \quad 0 \le i \le 3, 0 \le j \le 1,$$

we can easily show that G is isomorphic to the group of the square bipyramid.

More generally, for the regular polygon of n sides ($n > 2$), let x denote a rotation clockwise through $2\pi/n$ and y a (plane) reflection in the line containing the center of the polygon and the midpoint of one (selected) side. Then one can easily show

$$x^n = e, \qquad y^2 = e, \quad \text{and} \quad yx = x^{-1}y.$$

The group thus formed is called the *dihedral group of order 2n.* Its elements are all of the products of the form $x^i y^j$ with $0 \le i \le n-1$ and $0 \le j \le 1$. It is denoted by D_n; note that $|D_n| = 2n$. The potential confusion of the designation D_n with our notation for "stabilizer of point n under group D" will be avoided because we shall use D_n only to refer to the abstract dihedral group of order $2n$ and shall continue to use other letters when we speak of stabilizers.

When we are not numbering vertices as in figure 6, no confusion will result if we write 1 rather than e for the group identity, as we did in section 6. Hence we may hereafter assume that G *always denotes a group and 1 its identity.*

Now let p be any prime greater than 2 and suppose that G is a group of order $2p$. Then G might be isomorphic to Z_{2p} or to D_p. We show that in fact these are the only possibilities:

(7.2) THEOREM. If p is a prime greater than 2 and $|G| = 2p$, then either $G \cong Z_{2p}$ or $G \cong D_p$.

PROOF. From (6.8), s_p is a divisor of 2, and p divides $(s_p - 1)$. Hence $s_p = 1$, and by (6.6), G has a normal Sylow p-subgroup S_p. That S_p must be cyclic is exercise 7.5; we write $S_p = \{1, x, x^2, \ldots, x^{p-1}\}$. Also from (6.8), we find that s_2 must be 1 or p. If $s_2 = 1$, then G is cyclic as in our discussion of groups of order 35. Hence suppose that G has p Sylow 2-subgroups and let $\{1, y\}$ be one of them. Now if $\{1, z\}$ is a Sylow 2-subgroup distinct from $\{1, y\}$, then $z \ne y$. Thus G must have p distinct elements of order 2. But G has $p-1$ distinct elements of order p (namely, $x, x^2, \ldots, x^{p-1}$) and an identity, so we have accounted for all of the $2p$ elements. We thus conclude that xy must either be one of the p elements of order 2 or be one of the powers of x. If $xy = x^k$ for some k, then $y = x^{k-1}$. By exercise 3.10, the order of x^{k-1} must be a divisor of the order of S_p, hence 1 or p. But y has order 2, which gives a contradiction. We conclude that xy has order 2, that is, $(xy)^2 = 1$. From $xyxy = 1$ we have (using the fact that $y = y^{-1}$)

$$yx = x^{-1}y^{-1} = x^{-1}y,$$

and thus G has precisely the same elements and multiplication table as D_p. This completes the proof.

What we have done with D_n is to express this group in terms of *generators and relations.* To do so, we first specify a set of elements, called the *generators,* with the property that every element of the group can be written as a product of these generators and their powers. In addition, we give a list of equations, called the *relations,* that completely specify the group operation in the sense that the entire table is determined by those equations. For D_n our generators are x and y, and our relations are $x^n = 1$, $y^2 = 1$, and $yx = x^{-1}y$. We denote D_n in terms of generators and relations by

$$D_n = \langle x, y : x^n = 1, y^2 = 1, yx = x^{-1}y \rangle. \tag{7.3}$$

For brevity we may combine the first two relations as $x^n = y^2 = 1$. In the light of the last part of the proof of (7.2), we could have given $(xy)^2 = 1$ in place of the relation $yx = x^{-1}y$ in (7.3) since either of these two equations can be derived from the other.

As further examples of groups written in terms of generators and relations we have

$$\begin{cases} Z_k = \langle x : x^k = 1 \rangle, \\ V_4 = \langle x, y : x^2 = y^2 = 1, yx = xy \rangle. \end{cases} \tag{7.4}$$

The notation $\langle x \rangle$ stands for Z_∞ when no other context is clear. When some particular group is under consideration, $\langle x \rangle$ means (as in exercise 3.10) the subgroup consisting of all powers of x; thus in the context of D_n as denoted in (7.3), $\langle x \rangle$ would mean $\{1, x, x^2, x^3\}$.

An expression like (7.3) or one of (7.4) is called a *presentation* of the group in terms of generators and relations.

Since the generators and relations completely specify a group, in order to produce a homomorphism ϕ, one need only define function ϕ on the generators of the group and verify that the given relations are preserved. For example,

$$\phi(x) = x^{-1}, \qquad \phi(y) = y$$

defines a homomorphism on D_n since, first, the preservation of products and powers by ϕ determines what $\phi(z)$ must be for each $z \in D_n$, and second, ϕ preserves the relations in (7.3) as follows:

$$\begin{aligned} \phi(x^n) &= \phi(x)^n && \text{since } \phi \text{ must preserve powers} \\ &= (x^{-1})^n && \text{by definition of } \phi \\ &= 1 = \phi(1) && \text{by (4.1),} \end{aligned}$$

and similarly,

$$\begin{aligned} \phi(y^2) &= \phi(y)^2 = \phi(1), \\ \phi(yx) &= \phi(y)\phi(x) = yx^{-1} = (xy^{-1})^{-1} = (xy)^{-1} \\ &= xy = \phi(x^{-1})\phi(y) = \phi(x^{-1}y). \end{aligned}$$

The fact that a homomorphism is specified by its action on the generators alone is analogous to the result in linear algebra that a linear transformation is completely specified by its action on the elements of a basis.

A word of caution about generators and relations is in order. A presentation like (7.3) is a convenient form with which to work provided that one knows that such a group exists. However, just writing down a presentation in terms of generators and relations is no guarantee that it defines a group. By convention, when we write the relation $x^k = 1$ in a presentation, we mean that the elements $x, x^2, \ldots, x^{k-1}$ are all different from 1, that is, that the order of x is k. Now consider the innocuous presentation

$$\langle x, y : x^3 = y^3 = 1, yx = x^2y \rangle.$$

With a bit of patience one can produce a table for this construction, using the elements

$$1, x, x^2, y, y^2, xy, x^2y, xy^2, x^2y^2,$$

and erroneously conclude that the presentation gives a group of order 9. The fact is that the relations are contradictory, as we shall now show. Given $yx = x^2y$, we have $(yx)^{-1} = (x^2y)^{-1}$, whence $x^{-1}y^{-1} = y^{-1}x^{-2}$, which is $x^2y^2 = y^2x$. But

$$\begin{aligned} y^2x &= y(yx) = y(x^2y) = (yx)(xy) \\ &= (x^2y)(xy) = x^2(yx)y = x^2(x^2y)y \\ &= xy^2. \end{aligned}$$

Hence $x^2y^2 = xy^2$, which by cancellation gives $x = 1$, in contradiction to the relation $x^3 = 1$.

Another group that is important for applications in mathematics and in the physical sciences is the *quaternion group* Q_2 given by

$$Q_2 = \langle x, y : x^4 = 1, x^2 = y^2, yx = x^{-1}y \rangle.$$

Observe that from $x^4 = 1$ and $x^2 = y^2$ we can conclude that

$$xy^2 = x^3, \qquad y^3 = x^2y, \quad \text{and} \quad y^4 = 1;$$

hence we know that

$$xy^3 = x(x^2y) = x^3y,$$

and of all products of the form $x^a y^b$ we need consider only those with $a = 0, 1, 2, 3$ and $b = 0, 1$. This observation gives us eight elements

(7.5) $$1, x, x^2, x^3, y, xy, x^2y, x^3y.$$

Moreover, since $yx = x^{-1}y$, a power of x on the right may always be "crossed over" a y on the left, that is, the powers of x may be gathered at the left of a product, leaving the powers of y on the right. Hence (7.5) accounts for all elements of Q_2, and $|Q_2| = 8$.

A different characterization of Q_2, more usual in physical science applications, is given in exercise 7.9.

The final family of groups we shall consider in this section is those consisting of all possible permutations of a set of n points; such a group is denoted S_n and is called the *symmetric group of degree n.* I state without proof:

(7.6) THEOREM. $|S_n| = n!$ (where $n!$ denotes n-factorial).

(7.7) THEOREM. S_n is generated by the elements

$$(12), (13), \ldots, (1n).$$

Clearly, the square-bipyramid group in section 1 is a subgroup of S_6, and the symmetries of the square form a subgroup of S_4.

A permutation of the point set $\{1, 2, \ldots, n\}$ may be written in the useful form

$$\begin{pmatrix} 1 & 2 & \ldots & n \\ a_1 & a_2 & \ldots & a_n \end{pmatrix}$$

in which the indication is that $1 \to a_1$, $2 \to a_2, \ldots, n \to a_n$. Of course $a_1, \ldots, a_n$ is just a scrambling of the list $1, \ldots, n$. Now any permutation may be effected by a sequence of transpositions of adjacent elements in a list of the form $b_1, \ldots, b_n$. For example, the permutation written as (253) in section 1 may now be expressed (in

S_5) as

$$\begin{pmatrix} 1 & 2 & 3 & 4 & 5 \\ 1 & 5 & 2 & 4 & 3 \end{pmatrix}$$

and may be accomplished by the successive transpositions

$$\begin{array}{ccccc} 1 & 2 & 3 & 4 & 5 \\ & & \downarrow & (45) & \\ 1 & 2 & 3 & 5 & 4 \\ & & \downarrow & (35) & \\ 1 & 2 & 5 & 3 & 4 \\ & & \downarrow & (25) & \\ 1 & 5 & 2 & 3 & 4 \\ & & \downarrow & (34) & \\ 1 & 5 & 2 & 4 & 3 \end{array}$$

That is, $(253) = (45)(35)(25)(34)$, which may be verified as described in section 1. It is also true that $(253) = (25)(23)$, so such a decomposition into cycles of length 2 is not unique. However:

(7.8) THEOREM. Let $g \in S_n$; if g may be decomposed into an even number of cycles of length 2, then every such decomposition of g has an even number of cycles (and we call g an *even permutation*). Similarly, if one such decomposition of $h \in S_n$ has an odd number of cycles of length 2, then every such decomposition of h has an odd number of cycles (and we call h an *odd permutation*).

(7.9) THEOREM. The set of even permutations in S_n forms a normal subgroup of S_n, which we call the *alternating group of degree n* and denote by A_n. Moreover, $[S_n : A_n] = 2$ for every $n \geq 2$.

I omit the proofs here; the interested reader may consult, for example, Fraleigh [**6**], pp. 47–50.

For $n \geq 3$, S_n is nonabelian, and for $n > 3$, A_n is nonabelian. It follows from (7.9) and (7.6) that $|A_n| = n!/2$. Moreover, one can show that the tetrahedral group of exercise 1.7 is isomorphic to A_4 (see exercise 7.6). Two other useful results are:

(7.10) THEOREM. If $n \geq 5$, then A_n is simple.

(7.11) THEOREM. The alternating group A_n is generated by the elements (123), (124), ..., (12n).

The group A_5, of order 60, is the smallest nonabelian simple group. Its Sylow subgroups have orders 4, 3, and 5, and by (6.4), A_5 also has at least one subgroup of order 2. The remaining divisors of 60 are 6, 10, 15, 20, and 30; we shall show that A_5 has subgroups of order 6 and 10 but not of the other three orders. First we shall prove the following result, which applies to many finite groups:

(7.12) PROPOSITION. If $H \leq G$ and $[G:H]=2$, then $H \trianglelefteq G$.

PROOF. If $x \in G$ and $x \notin H$, then $G = H \cup Hx$. Let $g \in G$; we must show that $g^{-1}Hg = H$. If $g \in H$, the conclusion follows by closure. Otherwise, there exists an $h \in H$ such that $g = hx$. If $g^{-1}Hg \neq H$, there must be some $a \in H$ with $g^{-1}ag \notin H$, whence $(hx)^{-1}a(hx) = bx$ for some $b \in H$. Now we have $x^{-1}h^{-1}ahx = bx$, and by cancellation (exercise 2.8), $x^{-1}h^{-1}ah = b$; but then $x = h^{-1}ahb^{-1} \in H$, a contradiction. (For another proof see exercise 7.18.)

Now we can show, using (7.11), that A_5 has no subgroup of order 30. Suppose $H \leq A_5$ and $|H| = 30$; then $[A_5:H]=2$ and so $H \trianglelefteq A_5$. By the converse of (5.5), there is a homomorphism $\eta: A_5 \to A_5/H$. Since each of the generators (123), (124), (125) has order 3 (see exercise 3.12), it must be in the kernel H of η. (See exercise 5.11.) But then H contains all the elements of A_5 and must equal A_5, which contradicts the assumption that $|H| = 30$.

Next we show that A_5 has subgroups of order 10. By the third Sylow theorem (6.8), s_5 must be 1 or 6. We can see that $s_5 \neq 1$ even without appealing to (7.10), as follows. A 5-cycle such as (12345) is an even permutation, and A_5 must therefore contain (12345) and (13245) together with their powers; thus we account for eight elements of order 5, too many for a unique Sylow 5-subgroup. Hence there are six Sylow 5-subgroups, and the normalizer of any given Sylow 5-subgroup must have order 10 by (6.7); this normalizer is a subgroup by exercise 5.1.

If A_5 has a subgroup H of order 15, then H has a Sylow 5-subgroup K, and K must also be a Sylow 5-subgroup of A_5. But by exercise 6.1, H is cyclic, so $K \trianglelefteq H$; hence H, a subgroup of order 15, is contained in the normalizer of K, which has order 10, a contradiction.

A similar argument shows that A_5 has no subgroup of order 20. If H were such a subgroup, then by (6.8), H would have a normal

Sylow 5-subgroup, and hence H would be contained in the normalizer of a Sylow 5-subgroup of A_5, which normalizer has order 10.

Finally, we show that A_5 does have subgroups of order 6. By (6.8), the number s_3 of Sylow 3-subgroups must be 1, 4, or 10. Since (123) and (124) generate distinct subgroups of order 3, we know that $s_3 \neq 1$. If $s_3 = 4$, then by (6.7), the normalizer of any Sylow 3-subgroup would have order 15, and we know that A_5 has no subgroup of this order. Hence $s_3 = 10$, and the normalizers of the Sylow 3-subgroups have order 6.

Incidentally, the reasoning above verifies (7.10) for the case $n = 5$.

At this point I conclude the presentation of the rudiments of abstract group theory. From time to time in the following sections, more group-theoretic results will be introduced as they are needed and when some of their applications can more easily be seen.

EXERCISES

7.1. Let k be any integer different from 0 and ± 1. Show that $\phi_k(x^i) = x^{ki}$ (for each integer i) defines an *iso*morphism of Z_∞ onto a proper subgroup of itself. Hence ϕ_k is *not* an *auto*morphism of Z_∞, although if $k = \pm 1$, we would have gotten an automorphism of Z_∞.

7.2. If H is a subgroup of a cyclic group, then H is cyclic.

7.3. If ϕ is a homomorphism on a cyclic group Z, then $\phi(Z)$ is cyclic.

7.4. Show that every cyclic group is abelian.

• *7.5.* Prove that if G has prime order, then G is cyclic.

7.6. Observe that any 3-cycle (abc) may be decomposed as $(ab)(ac)$ and hence is an even permutation. Then show that the elements of the tetrahedral group found in exercise 1.7 are precisely those of A_4.

7.7. Show that the dihedral group D_6 of order 12 is not isomorphic to A_4. (*Hint*: Use exercises 1.7, 3.12, and 5.11.)

7.8. Show that the symmetric group S_3 is nonabelian and hence is isomorphic to D_3. Why may we then speak of *the* nonabelian group of order 6?

7.9. The quaternion group Q_2 is often written in applications as $\{\pm 1, \pm i, \pm j, \pm k\}$, where $i^2 = j^2 = k^2 = -1$, $ij = k$, $jk = i$, and $ki = j$. Find an isomorphism between this group and Q_2 as given in the text preceding (7.5). (*Note*: The comments on defining a function only on the generators and then checking that relations are preserved apply here.)

• *7.10.* If G is a group, then the set of automorphisms of G forms a group $\mathfrak{A}(G)$ called the *automorphism group* of G, using the operation of composition of functions, that is, for σ, $\tau \in \mathfrak{A}(G)$, we take $(\sigma\tau)(g) = \sigma(\tau(g))$ for each $g \in G$. Find $\mathfrak{A}(Z_4)$ and $\mathfrak{A}(V_4)$, that is, identify them as abstract groups.

7.11. Continuing exercise 7.10, find $\mathfrak{A}(D_4)$.

7.12. In the notation of exercises 3.4 through 3.6, consider $K = \langle x, r^2 \rangle$. By finding what r^2 does to the point 1, then what x and x^2 do to the points 1 and 1^{r^2}, then what r^2 does to the points considered thus far, etc., show that the orbit of 1 under K consists of four points. What is $|K|$? Use the information just found together with the first part of exercise 5.10 to list the elements of the normal Sylow 2-subgroup of K. (*Hint*: It must contain r^2, which has order 2, since it is the unique Sylow 2-subgroup.)

7.13. Without writing out a list of the elements of K, but by using the Sylow theorems, find out how many Sylow 3-subgroups the group K of exercise 7.12 has. (*Hint*: Observe that the stabilizer G_1 has order 3. Then prove that K has *no* element of order 6.)

7.14. If the mapping $\sigma: x \to x^{-1}$ (for $x \in G$) is a homomorphism of G, prove that σ is an automorphism of G and that G is abelian.

7.15. Let $\mathfrak{I}(G)$ denote the inner automorphisms of G; prove that $\mathfrak{I}(G)$ is a normal subgroup of $\mathfrak{A}(G)$. Then let a function $\sigma: G \to \mathfrak{I}(G)$ be defined by $\sigma(g) = \phi_g$, where $\phi_g(x) = gxg^{-1}$ for each $x \in G$. (*Note*: ϕ_g conjugates x by g^{-1}.) Prove that σ is a homomorphism. Explain why σ is onto. Determine the kernel of σ. When is G isomorphic to $\mathfrak{I}(G)$?

• *7.16.* Prove that there are only two abstract groups of order 4. (*Hint*: If $|G| = 4$ and G is not cyclic, explain why G must have two distinct elements x and y of order 2. What can xy and yx be?)

7.17. Prove that there are exactly five nonisomorphic groups of order 8. (*Hint*: Three of them are Z_8, D_4, and Q_2, and the other two are abelian.)

7.18. Use exercise 4.9 to give an alternative (and shorter) proof of (7.12).

Part Two
REPRESENTATIONS

Section 8
Matrix Groups

Representation theory is concerned with using matrices in place of the elements of an abstract group or, less generally, of a group arising from geometric symmetries. In order to construct homomorphisms carrying group elements to matrices, one first needs to know when sets of matrices form groups with respect to the operation of matrix multiplication.

From here on, I assume that the reader is familiar with the most basic ideas of linear algebra: vectors, linear transformations, matrices, and determinants. Since most of the concepts needed will be recalled as they arise, I shall not digress here to a general review of elementary linear algebra.

For any positive integer n, we may consider the n-by-n matrices with entries from the field $\mathbf{C}$ of complex numbers. Recall that the identity matrix I_n has the property that for any other matrix A of the same size, $AI_n = I_nA = A$; hence I_n can function as an identity with respect to matrix multiplication. Recall also that a matrix A has an inverse A^{-1} with the property that $AA^{-1} = A^{-1}A = I_n$ if and only if A has nonzero determinant (that is, A is *nonsingular*). Further, since the determinant of a matrix product is the product of the determinants, the product of two nonsingular matrices is again nonsingular. These considerations lead us to:

(8.1) PROPOSITION. The set of nonsingular n-by-n matrices with entries from the complex numbers $\mathbf{C}$ forms a group $GL(n, \mathbf{C})$ under the operation of matrix multiplication. Those matrices having entries only from the real numbers

R form a subgroup $GL(n, \mathbf{R})$ of $GL(n, \mathbf{C})$. Either of these groups may be called the *general linear group of degree n;* when confusion may result, we may add *over the complex* (or *real*) *numbers.* (The term has its origin in linear algebra, where a nonsingular matrix specifies a linear transformation of an n-dimensional vector space with respect to a particular basis.)

Any subgroup of $GL(n, \mathbf{C})$ is called a *matrix group.* Such a group may be finite or infinite; $GL(n, \mathbf{C})$ and $GL(n, \mathbf{R})$ are both infinite. As an example of a finite matrix group, consider the four matrices

$$I = \begin{bmatrix} 1 & 0 \\ 0 & 1 \end{bmatrix}, \quad A = \begin{bmatrix} 0 & 1 \\ 1 & 0 \end{bmatrix}, \quad B = \begin{bmatrix} -1 & 0 \\ 0 & -1 \end{bmatrix}, \quad AB = \begin{bmatrix} 0 & -1 \\ -1 & 0 \end{bmatrix}.$$

It is easy to check that these four matrices form a group isomorphic to V_4. An example of a cyclic matrix group of order 8 is the powers of

$$B = \begin{bmatrix} 0 & -1 \\ i & 0 \end{bmatrix},$$

where $i = \sqrt{-1}$, as usual.

If $X = \begin{bmatrix} 1 & 1 \\ 0 & 1 \end{bmatrix}$, then $X^n = \begin{bmatrix} 1 & n \\ 0 & 1 \end{bmatrix}$ for any integer n, and thus $\langle X \rangle$ is an infinite cyclic group. It is isomorphic to the group consisting of the powers of

$$Y = \begin{bmatrix} 1 & i \\ 0 & 1 \end{bmatrix}.$$

(See exercise 8.2.)

An important subgroup of $GL(n, \mathbf{C})$ is the group $SL(n, \mathbf{C})$ consisting of the n-by-n matrices with determinant 1 and with complex entries. The corresponding subgroup of $GL(n, \mathbf{R})$ is denoted by $SL(n, \mathbf{R})$. Either of these groups is called the *special linear group* of degree n. Another subgroup of interest falls between the general linear and special linear groups; it is the subgroup consisting of the matrices of determinant ± 1.

Matrix groups, or correspondingly, groups of linear transformations under composition, give rise to the entire theory of group representations.

EXERCISES

8.1. Show that for $n=1$, $GL(n, \mathbf{R}) \trianglelefteq GL(n, \mathbf{C})$. Then show that normality does *not* hold for $n=2$, by considering $Y^{-1}XY$, where

$$X=\begin{bmatrix} 1 & 2 \\ 3 & 1 \end{bmatrix} \quad \text{and} \quad Y=\begin{bmatrix} 1 & i \\ 0 & 1 \end{bmatrix}.$$

8.2. Let

$$Y=\begin{bmatrix} 1 & i \\ 0 & 1 \end{bmatrix};$$

compute Y^2 and Y^3 and then exhibit two distinct isomorphisms between $\langle Y \rangle$ and the infinite cyclic group $\langle X \rangle$.

8.3. Prove that $SL(n, \mathbf{C}) \trianglelefteq GL(n, \mathbf{C})$. (*Hint*: Consider determinants.) Of course, $SL(n, \mathbf{R}) \trianglelefteq GL(n, \mathbf{R})$ also.

• *8.4.* Let ζ be a (complex) primitive k^{th} root of 1, that is, ζ is a complex number with $\zeta^k=1$ and with $\zeta^m \neq 1$ if $1 \leq m < k$. Let G consist of the matrices in $GL(n, \mathbf{C})$ whose determinants are powers of ζ. Prove that $G \trianglelefteq GL(n, \mathbf{C})$.

8.5. Let

$$A=\begin{bmatrix} 0 & 1 & 0 \\ 0 & 0 & 1 \\ 1 & 0 & 0 \end{bmatrix} \quad \text{and} \quad B=\begin{bmatrix} 0 & -1 & 0 \\ 0 & 0 & 1 \\ 1 & 0 & 0 \end{bmatrix}.$$

Show that A and B generate cyclic groups of order 3 and 6, respectively.

8.6. Let A be the same as in the preceding exercise, and let $C=(-1)I_3$. Show that $C^2=I_3$ and that $AC=CA$. Then without performing any other matrix multiplications, show that $\langle A, C \rangle$ is cyclic of order 6.

8.7. Prove that $GL(n, \mathbf{C})$ forms a group under the operation of matrix multiplication.

Section 9

Group Representations

Let G be an arbitrary group and G^* a matrix group, that is, a subgroup of $GL(n, \mathbf{C})$. A homomorphism $T: G \to G^*$ is called a *representation of* G; the integer n is the *degree* or *dimension* of T. Thus a representation of G replaces the elements of G by n-by-n matrices, and the group multiplication is replaced by matrix multiplication.

For example, let G be the cyclic group of order 2 (here we consider G an abstract group), and write $G = \{1, x\}$. Then

$$T(x) = \begin{bmatrix} 0 & 1 \\ 1 & 0 \end{bmatrix}, \qquad T(1) = \begin{bmatrix} 1 & 0 \\ 0 & 1 \end{bmatrix}$$

is a representation of G; the dimension of T is 2. For a one-dimensional representation of G, let $U(x) = [-1]$ and $U(1) = [1]$. Note that a one-dimensional representation is a function whose values are 1-by-1 matrices, which are distinguished in form from real or complex numbers as such.

We have defined a representation to be a particular kind of function between a group in which we are interested and a matrix group. The value of representations is primarily in that computations with matrices are relatively easy to handle (whereas group operations can be messy) and that many theorems about matrices illuminate the theory of groups by means of representations.

Note that if 1 is the identity of G and T is any representation of G, then $T(1) = I_n$ (where n is the dimension of T) by (4.1). Moreover, if $g \in G$, then $T(g^{-1}) = T(g)^{-1}$, and for any integer k, $T(g^k) = T(g)^k$, the kth power of the matrix $T(g)$. (If k is negative, the kth power of a matrix is, of course, the $-k^{\text{th}}$ power of its inverse.) Thus to specify a representation, we need only give the values of T on a set of generators, as discussed in section 7.

We consider next some representations of the groups of order 4 (recall from exercise 7.16 that there are exactly two abstract groups of order 4). Let $G=\langle x\rangle$ be cyclic of order 4, and let

$$T(x)=\begin{bmatrix}1&0&0\\0&0&-1\\0&1&0\end{bmatrix}. \tag{9.1}$$

Since $T(x)^4=I_3$, the defining relation $x^4=1$ of G is preserved by T, and T is a representation of G. For the same group, let

$$U(x)=\begin{bmatrix}0&1\\1&0\end{bmatrix};$$

Then $U(x)^4=I_2$, and U is a representation of G. The fact that x^2 is in the kernel of U means that $U(x^2)=U(1)$ (although $x^2\neq 1$) and, as is readily verified, that $U(x^3)=U(x)$. Thus U uses only two matrices to represent the group G of order 4, whereas T uses four matrices (see exercise 9.1).

Now let $G\cong V_4$; then as in section 7, G may be written in terms of generators and relations as

$$\langle x, y: x^2=y^2=1,\ yx=xy\rangle.$$

Let $T(x)=\begin{bmatrix}0&1\\1&0\end{bmatrix}$ and $T(y)=\begin{bmatrix}-1&0\\0&-1\end{bmatrix}$; to check that the defining relations for G are preserved, we observe that $T(x)^2=T(y)^2=I_2$ and that

$$\begin{aligned}T(xy)=T(x)T(y)&=\begin{bmatrix}0&1\\1&0\end{bmatrix}\begin{bmatrix}-1&0\\0&-1\end{bmatrix}=\begin{bmatrix}0&-1\\-1&0\end{bmatrix}\\&=T(y)T(x)=T(yx).\end{aligned}$$

Hence T is a representation of G. For the same group let $U(x)=\begin{bmatrix}0&1\\1&0\end{bmatrix}$ and $U(y)=I_2$. Then U is a representation of G with kernel $\{1, y\}$.

In these examples for the groups of order 4, the kernel of T in each case is the trivial subgroup $\{1\}$, but the kernel of U has order 2. Thus each T is an isomorphism, and neither U is an isomorphism. A representation that is an isomorphism is a 1:1 function and so has a distinct matrix for each element of the group; such a representation is called *faithful.*

Now we shall find all of the *one-dimensional* representations of $S_3=\langle x, y: x^3=y^2=1,\ yx=x^2y\rangle$. Here we have given the presentation for D_3; by theorem (7.2) the only groups of order 6 are Z_6 and

D_3, but $|S_3| = 6$ and S_3 is easily shown to be nonabelian, so $S_3 \cong D_3$. Suppose that T is a one-dimensional representation of S_3. Then $T(y)$ must be a 1-by-1 matrix whose square is the identity matrix I_1; thus $T(y) = [\pm 1]$. Similarly, if $T(x) = [\zeta]$, then we need $T(x)^3 = T(1)$, whence $\zeta^3 = 1$. Now suppose we let ζ be a *primitive* cube root of 1 (see exercise 8.4), that is,

$$\zeta = -\frac{1}{2} \pm i\frac{\sqrt{3}}{2}.$$

If $T(x) = [\zeta]$ and $T(y) = [1]$, then $T(yx) = [\zeta]$ but $T(x^2y) = [\zeta^2]$. Since $\zeta \neq \zeta^2$, T does not preserve the relation $yx = x^2y$ and so is not a representation of S_3. Similarly, if $T(x) = [\zeta]$ and $T(y) = [-1]$, then $T(yx) \neq T(x^2y)$. Thus our choice of ζ to be a *primitive* cube root of 1 cannot lead to a one-dimensional representation of S_3. But, as we observed above, if $T(x) = [\zeta]$, then ζ^3 must be 1, so the only choice remaining is $\zeta = 1$. If $T(x) = [1]$ and $T(y) = [\pm 1]$, then $T(yx) = [\pm 1] = T(x^2y)$; thus for either sign in $T(y)$ we have a representation of S_3 and, in fact, we have shown that these two are the *only* one-dimensional representations of S_3.

Observe that one of the two representations we just found merely assigns the matrix $[1]$ to every element of S_3. In general, if G is any group and n is any positive integer, then the function

$$T(g) = I_n \quad \text{for every } g \in G$$

is a representation of G. It is called the *trivial representation of degree n*, but the triviality is in the construction of this representation—it is not trivial in the sense of being unimportant!

Now an n-by-n matrix A determines a linear transformation on a vector space of dimension n. Specifically, if $A \in GL(n, \mathbf{R})$, then A gives a linear transformation on the Euclidean space $\mathbf{R}^n$, and if $A \in GL(n, \mathbf{C})$, then the linear transformation given by A is one on the space $\mathbf{C}^n$ of n-dimensional vectors with complex entries. We shall speak correspondingly of a representation of a group G *over the real numbers* $\mathbf{R}$ or *over the complex numbers* $\mathbf{C}$. We say that $\mathbf{R}^n$ or $\mathbf{C}^n$ *affords* a representation T of a group G if for each $g \in G$, $T(g)$ is in $GL(n, \mathbf{R})$ or $GL(n, \mathbf{C})$, respectively.

By considering the Euclidean spaces $\mathbf{R}^2$ (the plane) and $\mathbf{R}^3$, we can use our geometric examples from earlier sections to obtain matrix representations of the abstract groups to which they are isomorphic. First, however, some comment on notation will be needed. If X is an n-dimensional *column* vector and A an n-by-n

matrix, then from linear algebra we recall that AX is the vector to which the transformation determined by A carries X. Now consider the transposes X^t and A^t (the superscript ^{t} denotes "transpose"). In particular, X^t is a *row* vector. Now

$$(AX)^t = X^t A^t;$$

hence $X^t A^t$ is the row vector to which the transformation determined by A^t carries X^t. The reader most likely used column vectors in linear algebra; in representation theory it is more convenient to employ row vectors for two reasons. The first and less important is that it maintains consistency with our notation in (2.4) for a group acting on a set. The more crucial consideration is that we have interpreted the product xy for x and y in a group G to mean "x followed by y"; thus if T is a representation of G, we want $T(xy) = T(x)T(y)$ with the right-hand side meaning "$T(x)$ followed by $T(y)$." If we use *column* vectors X, we must define a representation T to be an *antihomomorphism*, that is,

$$T(xy) = T(y)T(x),$$

so that $T(xy)X = T(y)(T(x)X)$, but if we use *row* vectors Y, then

$$Y\,T(xy) = Y\,T(x)T(y)$$

is consistent with our definition of a representation as a homomorphism. These notational considerations should be clarified by the following example.

Consider the group D_4 of symmetries of the square in the Euclidean plane $\mathbf{R}^2$; we may write

$$D_4 = \langle r, c : r^4 = c^2 = 1,\ cr = r^3 c\rangle$$

as in section 7. Let r represent a rotation through 90° counterclockwise; thus if (x, y) is a row vector representing a point in $\mathbf{R}^2$,

$$(x, y)\begin{bmatrix} 0 & 1 \\ -1 & 0 \end{bmatrix} = (-y, x)$$

is the linear transformation that rotates the vector as required. Now c represents a reflection, say in the horizontal axis of figure 6 (section 7), so we want the linear transformation

$$(x, y)\begin{bmatrix} 1 & 0 \\ 0 & -1 \end{bmatrix} = (x, -y).$$

Thus we let

$$(9.2)\qquad T(r)=\begin{bmatrix}0 & 1\\ -1 & 0\end{bmatrix}\quad\text{and}\quad T(c)=\begin{bmatrix}1 & 0\\ 0 & -1\end{bmatrix};$$

one can easily verify algebraically that the homomorphism T so determined preserves the relations for D_4 (see exercise 9.3). Alternatively, we may simply observe that because the linear transformations we have found here describe the symmetries geometrically, it is clear that we have a faithful representation of D_4.

If we had dealt with column vectors rather than row vectors, we would have had the same $T(c)$, but $T(r)$ would be $\begin{bmatrix}0 & -1\\ 1 & 0\end{bmatrix}$ in order to have $T(r)\begin{bmatrix}x\\ y\end{bmatrix}=\begin{bmatrix}-y\\ x\end{bmatrix}$ as required. We could then have checked that the symmetry rc (r followed by c) was given by the linear transformation whose matrix is $\begin{bmatrix}0 & -1\\ -1 & 0\end{bmatrix}=T(c)T(r)$, and *not* by $\begin{bmatrix}0 & 1\\ 1 & 0\end{bmatrix}=T(r)T(c)$.

We proved in section 7 that D_4 is isomorphic to the square-bipyramid group of section 1. Hence we may consider the geometry of the square bipyramid in order to find another representation of D_4. Place the solid with its center at the origin in $\mathbf{R}^3$ and its central square in the XY-plane, as in figure 7. The rotation r of section 1 may be replaced (for convenience) by a *counterclockwise* rotation through 90° about the Z-axis; this motion is given by

$$(x, y, z)\begin{bmatrix}0 & 1 & 0\\ -1 & 0 & 0\\ 0 & 0 & 1\end{bmatrix}=(-y, x, z).$$

The motion c of section 1 may be taken as a rotation through 180° about the X-axis, which is given by

$$(x, y, z)\begin{bmatrix}1 & 0 & 0\\ 0 & -1 & 0\\ 0 & 0 & -1\end{bmatrix}=(x, -y, -z).$$

Thus $V(r)=\begin{bmatrix}0 & 1 & 0\\ -1 & 0 & 0\\ 0 & 0 & 1\end{bmatrix}$ and $V(c)=\begin{bmatrix}1 & 0 & 0\\ 0 & -1 & 0\\ 0 & 0 & -1\end{bmatrix}$ determines a representation of degree 3 of D_4. Note that each of these matrices

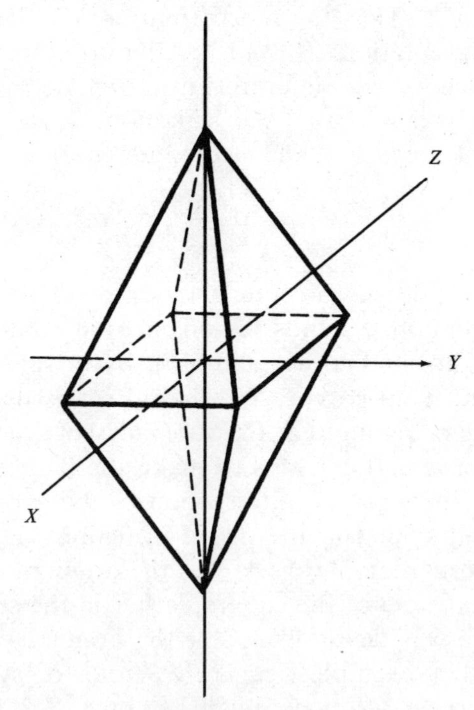

Figure 7

may be partitioned as

$$\left[\begin{array}{cc:c} a_{11} & a_{12} & 0 \\ a_{21} & a_{22} & 0 \\ \hdashline 0 & 0 & a_{33} \end{array}\right].$$

The product of any two such matrices again has this form (see exercise 9.4). More specifically, if we let $U(r)=[1]$ and $U(c)=[-1]$, then U is easily seen to be a one-dimensional representation of D_4. Indeed, if T is the two-dimensional representation obtained above from symmetries of the square, then

$$V(r)=\left[\begin{array}{cc:c} \multicolumn{2}{c:}{\multirow{2}{*}{$T(r)$}} & 0 \\ & & 0 \\ \hdashline 0 & 0 & U(r) \end{array}\right], \qquad V(c)=\left[\begin{array}{cc:c} \multicolumn{2}{c:}{\multirow{2}{*}{$T(c)$}} & 0 \\ & & 0 \\ \hdashline 0 & 0 & U(c) \end{array}\right],$$

and by virtue of exercise 9.4, we have

$$V(g)=\left[\begin{array}{cc:c} \multicolumn{2}{c:}{\multirow{2}{*}{$T(g)$}} & 0 \\ & & 0 \\ \hdashline 0 & 0 & U(g) \end{array}\right]$$

for every $g \in D_4$. Thus the representation V decomposes into two smaller representations T and U. Part of the problem of representation theory is to determine how such decompositions occur.

Conversely, if we have representations T and U of a group G, with m the degree of T and n the degree of U, then

$$V(g) = \begin{bmatrix} T(g) & \bar{0} \\ \bar{0} & U(g) \end{bmatrix} \tag{9.3}$$

for each $g \in G$, defines a representation of G, as is easy to see. In (9.3), the notation $\bar{0}$ stands for an m-by-n submatrix of zeros in the upper right-hand corner and an n-by-m submatrix of zeros in the lower left-hand corner. We shall use this notation frequently with the understanding that the submatrix of zeros has the proper number of rows and columns to make the large matrix square.

The placement of the intersection of the axes of symmetry of the square bipyramid at the origin in figure 7 is essential since a linear transformation always carries the origin of $\mathbf{R}^n$ to itself. Note that in the example of the square earlier in the section, the center of the square was tacitly placed at the origin.

For another example, recall the group of symmetries of the cube from section 3. For simplicity, we take G to include only rigid symmetries; thus $|G| = 24$. This time we take r to be a *clockwise* rotation (when seen from above) through 90° about the Z-axis. Then r may be represented by the cube of the matrix $V(r)$ from the preceding example (Why?), and we set

$$T(r) = \begin{bmatrix} 0 & -1 & 0 \\ 1 & 0 & 0 \\ 0 & 0 & 1 \end{bmatrix}. \tag{9.4}$$

Comparing figure 8 with figure 4 (section 3), we may assign coordinates to the vertices of the cube as follows:

Vertex (Fig. 4)	Coordinates (Fig. 8)
1	$(1, -1, 1)$
2	$(-1, -1, 1)$
3	$(-1, 1, 1)$
4	$(1, 1, 1)$
5	$(1, -1, -1)$
6	$(-1, -1, -1)$
7	$(-1, 1, -1)$
8	$(1, 1, -1)$

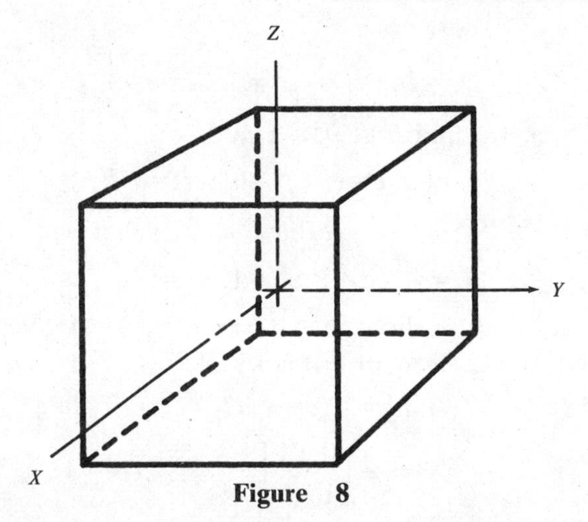

Figure 8

Then in terms of the vertices, r is the permutation (1234)(5678). Let $d=(245)(386)$; the motion d is described by the linear transformation

$$T(d)=\begin{bmatrix}0 & -1 & 0\\ 0 & 0 & -1\\ 1 & 0 & 0\end{bmatrix}. \tag{9.5}$$

With a little patience one can verify that G is generated by the two elements r and d; hence to define T on the rest of G we merely extend by powers and products as usual. It remains to show that T is indeed a homomorphism on G, but this fact is tedious to verify by calculation and easy to check by observation. Indeed, T is merely a specification of the linear transformations performed by the elements of G; thus the set of matrices

$$\{T(g): g\in G\} \tag{9.6}$$

forms a group that is isomorphic to G in the sense of being the same elements in a different notation, that is, (9.6) is the same abstract group as G.

Although in this example the matrix $T(r)$ has the form

$$\begin{bmatrix}A(r) & \bar{0}\\ \bar{0} & B(r)\end{bmatrix}$$

with $A(r)$ 2-by-2 and $B(r)$ 1-by-1, $T(d)$ does not have this form. We shall see later that it cannot be rewritten in such a form.

To introduce the last main point of this section, let us return to the group D_4, regarded as the symmetries of the square. If we write

$$D_4 = \langle r, c : r^4 = c^2 = 1, cr = r^3 c\rangle \tag{9.7}$$

and take $b = cr$, we find that $b^2 = 1$ and

$$br = (cr)r = (r^3 c)r = r^3(cr) = r^3 b;$$

moreover, $c = br^{-1} = br^3$, so we have shown that

$$D_4 = \langle r, b : r^4 = b^2 = 1, br = r^3 b\rangle. \tag{9.8}$$

In fact, one can easily show that $\sigma(r) = r$, $\sigma(c) = b$ gives rise to an isomorphism of D_4. Now in terms of (9.2),

$$T(b) = T(cr) = T(c)T(r)$$
$$= \begin{bmatrix} 0 & 1 \\ 1 & 0 \end{bmatrix}.$$

In view of the isomorphic presentations (9.7) and (9.8), we may define a representation T^* of D_4 by

$$T^*(r) = T(r), \quad T^*(c) = T(b), \tag{9.9}$$

bearing in mind, as usual, that we thereby mean that

$$T^*(r^i c^j) = T^*(r)^i T^*(c)^j$$

for $0 \le i \le 3$ and $0 \le j \le 1$.

Now T and T^* are different functions (they have different values on c, among other elements), but they were produced from a single geometric interpretation, in which b merely represents reflection in the line $x = y$, by exploiting the isomorphism σ. Thus in a very concrete sense, T and T^* are the same representation—they amount to the same set of linear transformations of $\mathbf{R}^2$. We need an explicit formulation of the statement that T and T^* are essentially the same representation, even though they are not identical as functions.

The idea of equivalence of representations is derived from that of linear transformations with a change of basis. Recall that two linear transformations L and L^* are *equivalent* if there is a change-of-basis for which $L^* \circ B = B \circ L$ (where $B \circ L$ means the composition of the functions, thus "L followed by B"). We illustrate this situation by the diagram

$$\begin{array}{ccc} V & \xrightarrow{L} & V \\ \big\downarrow{\scriptstyle B} & & \big\downarrow{\scriptstyle B} \\ V & \xrightarrow{L^*} & V \end{array} \tag{9.10}$$

in which the arrows indicate that the effect of applying L and then B to the vectors in V is the same as the effect of applying B and then L^*.

In terms of matrices, two matrices M and M^* are *equivalent* if there is a nonsingular matrix A such that $M^*A = AM$, or what is the same thing, if $A^{-1}M^*A = M$. If we write $L(X) = MX$ and $L^*(X) = M^*X$ (using column vectors X now that we are discussing linear algebra), the two ideas of equivalence are the same. The matrix A represents the change-of-basis B in the vector space V. Now we are ready to apply these ideas to representation theory:

(9.11) DEFINITION. Let T and U be n-dimensional representations of G; we say that T and U are *equivalent* if there is a nonsingular matrix A such that $AT(g) = U(g)A$ for *every* $g \in G$.

Note that in this definition *the same matrix* A must work simultaneously *for every* $g \in G$; thus A must be independent of the element g in the equation $AT(g) = U(g)A$.

To illustrate the definition, we return to T and T^* for D_4. In order to show that T and T^* are equivalent, we need a matrix

$$A = \begin{bmatrix} a & b \\ c & d \end{bmatrix}$$

with $ad - bc \neq 0$ such that $AT(g) = T^*(g)A$ for every $g \in D_4$. In particular, we need $AT(r) = T^*(r)A$, so

$$\begin{bmatrix} a & b \\ c & d \end{bmatrix}\begin{bmatrix} 0 & 1 \\ -1 & 0 \end{bmatrix} = \begin{bmatrix} 0 & 1 \\ -1 & 0 \end{bmatrix}\begin{bmatrix} a & b \\ c & d \end{bmatrix}$$

that is,

$$\begin{bmatrix} -b & a \\ -d & c \end{bmatrix} = \begin{bmatrix} c & d \\ -a & -b \end{bmatrix},$$

which gives $a = d$ and $b = -c$. Moreover, we want $AT(c) = T^*(c)A$, so

$$\begin{bmatrix} a & b \\ c & d \end{bmatrix}\begin{bmatrix} 1 & 0 \\ 0 & -1 \end{bmatrix} = \begin{bmatrix} 0 & 1 \\ 1 & 0 \end{bmatrix}\begin{bmatrix} a & b \\ c & d \end{bmatrix}$$

that is,

$$\begin{bmatrix} a & -b \\ c & -d \end{bmatrix} = \begin{bmatrix} c & d \\ a & b \end{bmatrix},$$

which yields $a = c$ and $b = -d$. Now for any constant $\lambda \neq 0$, $AT(g) = T^*(g)A$ if and only if $(\lambda A)T(g) = T^*(g)(\lambda A)$, so we have a free choice of one nonzero entry in A, say $a=1$. Then $A=\begin{bmatrix}1 & -1\\ 1 & 1\end{bmatrix}$, and one can easily verify that $AT(g)=T^*(g)A$ for every $g\in D_4$. Instead of making the arbitrary choice $a=1$, we could have chosen $a=1/\sqrt{2}$ to make A orthogonal.

We shall return to equivalence in later sections.

EXERCISES

9.1. Verify that T as given in (9.1) determines a cyclic matrix group of order 4.

• *9.2.* Prove that all of the one-dimensional representations of $V_4=\langle x, y : x^2=y^2=1, yx=xy\rangle$ are precisely those given by

$$\begin{aligned} T_1(x)&=[\ \ 1], & T_1(y)&=[\ \ 1];\\ T_2(x)&=[\ \ 1], & T_2(y)&=[-1];\\ T_3(x)&=[-1], & T_3(y)&=[\ \ 1];\\ T_4(x)&=[-1], & T_4(y)&=[-1]. \end{aligned}$$

9.3. Verify by taking matrix products that T as given in (9.2) determines a representation of D_4.

• *9.4.* Consider a family of matrices of the form

$$\begin{bmatrix}A & \bar{0}\\ \bar{0} & B\end{bmatrix},$$

where A is an n-by-n matrix, B is an m-by-m matrix, and the $\bar{0}$'s are taken as described in the text. Prove that the product of any two such matrices has the same form. If such a matrix is nonsingular, prove that its inverse has this same form.

9.5. Let V be the three-dimensional representation of the square-bipyramid group given in the text. Find $V(g)$ for each element of the group.

9.6. Show that T as given by (9.2) determines the same eight matrices as T^* as given by (9.9).

9.7. Given that

$$T(r)=\begin{bmatrix}0 & -1\\ 1 & 0\end{bmatrix}, \quad T(c)=\begin{bmatrix}0 & 1\\ 1 & 0\end{bmatrix}$$

and
$$U(r)=\begin{bmatrix}0 & -1 & 0 & 0\\ 1 & 0 & 0 & 0\\ 0 & 0 & 0 & -1\\ 0 & 0 & 1 & 0\end{bmatrix}, \quad U(c)=\begin{bmatrix}0 & 1 & 1 & 0\\ 1 & 0 & 0 & -1\\ 0 & 0 & 0 & 1\\ 0 & 0 & 1 & 0\end{bmatrix}$$

both determine representations for D_4 (expressed as in (9.7)), prove that U is equivalent to the representation

$$V(g) = \begin{bmatrix} T(g) & \bar{0} \\ \bar{0} & T(g) \end{bmatrix} \quad \text{for } g \in D_4.$$

9.8. Consider S_3 as the symmetries of an equilateral triangle in the plane $\mathbf{R}^2$. Use an argument similar to that preceding (9.2) and to that that produced the three-dimensional representation of the square-bipyramid group to find a two-dimensional representation of S_3.

9.9. Let $T(r) = [1]$, $T(c) = [-1]$ determine the representation found in the text for S_3, and let U be the trivial representation of S_3, where $S_3 = \langle r, c : r^3 = c^2 = 1, cr = r^2 c\rangle$. Prove that the representation found in exercise 9.8 is *not* equivalent to

$$V(g) = \begin{bmatrix} T(g) & 0 \\ 0 & U(g) \end{bmatrix}, \qquad g \in S_3.$$

(*Hint:* Postulate the existence of an A satisfying (9.11) for the two representations in question, and derive contradictory conditions on the entries a, b, c, d.)

9.10. Show that

$$T(g) = \begin{bmatrix} 0 & 1 \\ -1 & 0 \end{bmatrix}$$

determines a representation of $Z_4 = \langle g : g^4 = 1\rangle$ corresponding to a rotation of 90° in the plane. Then show that $U(g) = [-i]$ and $U^*(g) = [i]$ also determine representations of Z_4. Finally, show that T is equivalent to the representation

$$V(h) = \begin{bmatrix} U(h) & 0 \\ 0 & U^*(h) \end{bmatrix}, \qquad h \in Z_4.$$

(Since U and U^* involve complex numbers, the matrix A in (9.11) may be expected to contain complex numbers.)

Section 10
Regular Representations

This section introduces one of the most important representations of a group. Let us begin with a new action of a group G on the set of its own elements, an action that in fact specializes exercise 2.14 to subsets of G consisting of a single element.

(10.1) PROPOSITION. Let G be any group and for any x, $g \in G$, let

$$g^x = gx,$$

that is, x acts on the elements of G by multiplication on the right. Then this operation is an action of G on the set G.

PROOF. By the closure property in a group, g^x is always defined. To check conditions (2.4), we observe that

$$(g^x)^y = (gx)^y = gxy = g^{(xy)}$$

and

$$g^1 = g \cdot 1 = g.$$

We may also denote the action of right multiplication in functional notation by regarding each $x \in G$ as a function

$$(10.2) \qquad x_R : g \to gx \quad \text{for} \quad g \in G.$$

For action (10.2), the orbit of an element $g \in G$ under the group G is all of G, that is, $g^G = G$ in the notation of (2.5). To check this fact, we need only recall from (2.1)(c) that for any h, $g \in G$, there exists a unique $x \in G$ such that $gx = h$. In fact, given h and g, we take $x = g^{-1}h$. Moreover, if $g \in G$, then the stabilizer of g under G is the identity subgroup $\{1\}$ since $gx = g$ if and only if $x = 1$, by cancellation.

Now let G be a finite group of order n and write

$$G = \{x_1, x_2, \ldots, x_n\}.$$

We consider the action of a *fixed* element x_i on the set G. The result is a permutation of the elements x_j among themselves; for each j with $1 \le j \le n$, there exists a unique k with $1 \le k \le n$ such that

$$x_j x_i = x_k, \tag{10.3}$$

that is, $(x_i)_R : x_j \to x_k$. Equivalently, for each j there is a unique k such that

$$x_i = x_j^{-1} x_k. \tag{10.4}$$

Let $\mathbf{R}^n$ denote n-dimensional Euclidean space, as usual. To each x_j of G, with $1 \le j \le n$, we associate the unit *row* vector e_j having 1 as its jth component and 0 for each of the remaining components. Then, under this correspondence, (10.3) becomes

$$(x_i)_R : e_j \to e_k \tag{10.5}$$

for the unique k that we associated with each j. Formula (10.5) then specifies a linear transformation L on $\mathbf{R}^n$ by the customary method of defining the value of L on each member of the natural basis. The matrix M associated with L has a 1 in its (j, k)-entry if (10.5) holds for j and k; all other entries of M are 0. Note that in these two paragraphs we have held x_i fixed, though arbitrary; thus (10.5) induces a *different* linear transformation for *each* $i = 1, 2, \ldots, n$.

For an easy example, let $G = \{1, g, g^2\}$; we denote the elements by $x_1 = 1$, $x_2 = g$, and $x_3 = g^2$. Then the three instances of (10.5) become

$$(x_1)_R = 1_R : e_j \to e_j \quad \text{for} \quad j = 1, 2, 3;$$

$$(x_2)_R = g_R : \begin{cases} e_j \to e_{j+1} & \text{for} \quad j = 1, 2, \\ e_3 \to e_1; \end{cases}$$

$$(x_3)_R = g_R^2 : \begin{cases} e_1 \to e_3 \\ e_j \to e_{j-1} & \text{for} \quad j = 2, 3. \end{cases}$$

Hence the matrix $T(g)$ representing the linear transformation g_R must satisfy

$$e_1 T(g) = e_2, \qquad e_2 T(g) = e_3, \qquad e_3 T(g) = e_1.$$

Therefore,

$$T(g) = \begin{bmatrix} 0 & 1 & 0 \\ 0 & 0 & 1 \\ 1 & 0 & 0 \end{bmatrix}$$

and by similar reasoning,

$$T(g^2)=\begin{bmatrix} 0 & 0 & 1 \\ 1 & 0 & 0 \\ 0 & 1 & 0 \end{bmatrix}$$

and $T(1)=I_3$, the identity matrix. Clearly, $T(g^2)=T(g)^2$ and $T(g)^3=T(1)$, so T is indeed a representation of G.

The representation obtained from (10.3) and (10.5) for a finite group G is called the *right regular representation* of G. If we modify the discussion to use multiplication on the left (denoted by x_L) in (10.2), we obtain the *left regular representation* of G. For an abelian group, multiplication on the left is the same as multiplication on the right, so the two representations are identical.

For a nonabelian example, one naturally wants to choose the *smallest* nonabelian group in order to keep the matrices as small as possible; by exercises 7.5 and 7.16, we know that the smallest nonabelian group is of order 6, namely D_3, which has the presentation

$$D_3=\langle r, c : r^3=c^2=1, cr=r^2c\rangle.$$

Using the familiar notation for D_3 (rather than $x_1, \ldots, x_6$) we have

$$r_R : 1\to r, \qquad r_R : r \to r^2, \qquad r_R : r^2 \to 1,$$
$$r_R : c \to r^2c, \qquad r_R : rc \to c, \qquad r_R : r^2c \to rc.$$

We identify the elements $1, r, r^2, c, rc, r^2c$ with the basis vectors $e_1, \ldots, e_6$, respectively, and translate the above as

$$r_R : e_1 \to e_2, \qquad r_R : e_2 \to e_3, \qquad r_R : e_3 \to e_1,$$
$$r_R : e_4 \to e_6, \qquad r_R : e_5 \to e_4, \qquad r_R : e_6 \to e_5.$$

Hence the right regular representation for D_3 will have

$$T(r)=\begin{bmatrix} 0 & 1 & 0 & 0 & 0 & 0 \\ 0 & 0 & 1 & 0 & 0 & 0 \\ 1 & 0 & 0 & 0 & 0 & 0 \\ 0 & 0 & 0 & 0 & 0 & 1 \\ 0 & 0 & 0 & 1 & 0 & 0 \\ 0 & 0 & 0 & 0 & 1 & 0 \end{bmatrix}. \tag{10.6}$$

Since c_R may be summarized as

$$1 \to c, \qquad r \to rc, \qquad r^2 \to r^2c,$$
$$c \to 1, \qquad rc \to r, \qquad r^2c \to r^2,$$

the right regular representation for D_3 has

$$(10.7) \qquad T(c) = \begin{bmatrix} 0 & 0 & 0 & 1 & 0 & 0 \\ 0 & 0 & 0 & 0 & 1 & 0 \\ 0 & 0 & 0 & 0 & 0 & 1 \\ 1 & 0 & 0 & 0 & 0 & 0 \\ 0 & 1 & 0 & 0 & 0 & 0 \\ 0 & 0 & 1 & 0 & 0 & 0 \end{bmatrix}.$$

It is not difficult to repeat the argument above for r^2, rc, and r^2c and to verify that the resulting T is a representation by showing that it preserves the three relations in the presentation of D_3.

As is pointed out in the parallel though more sophisticated discussion of regular representations in Curtis and Reiner [**4**], pp. 33–34, this representation may be read off from a suitably arranged group table. Let $G = \{x_1, \ldots, x_n\}$; for each $i = 1, \ldots, n$, let $T(x_i)$ be the matrix of the linear transformation determined by (10.5). Since $T(x_i)$ merely permutes the vectors of the natural basis for $\mathbf{R}^n$, each $T(x_{i_i})$ will contain exactly one 1 in each row, exactly one 1 in each column, and zeros in the remaining $n^2 - n$ positions. Such a matrix is called a *permutation matrix.* In particular,

$$(10.8) \qquad \begin{cases} T(x_i) \text{ has a 1 as its } (j, k)\text{-entry} \\ \qquad \text{iff } (x_i)_R : e_j \to e_k \\ \qquad \text{iff } x_j x_i = x_k \\ \qquad \text{iff } x_i = x_j^{-1} x_k, \end{cases}$$

as pointed out earlier in this section.

Now (10.8) may be paraphrased by saying that the table for G must read, in part,

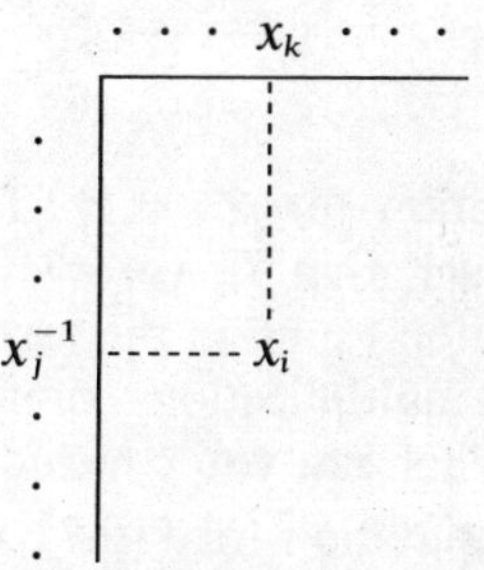

Thus if we prepare a group table whose *columns* are designated as $x_1, x_2, \ldots, x_n$ and whose *rows* are marked $x_1^{-1}, x_2^{-1}, \ldots, x_n^{-1}$, then:

(10.9) $\left\{ \begin{array}{l} T(x_i) \text{ has a 1 in its } (j, k)\text{-entry if and} \\ \text{only if the table described has } x_i \text{ in} \\ \text{row } j, \text{ column } k. \end{array} \right.$

In connection with (10.9), it is important to note that the group table in question is not arranged as tables have been in previous sections. This table has x_k^{-1} assigned to the kth row and x_k to the kth column; hence the identity element 1 appears in all of the entries on the main diagonal of the table. By (10.9), this means that $T(1)$ is the identity matrix I_n, as indeed we know it must be.

Of course there is no single order which must be followed in listing the elements of G as $x_1, \ldots, x_n$. A change in the order of elements in this list, however, merely reflects a permutation of the natural basis vectors e_j $(1 \leq j \leq n)$, which is a change of basis. In view of (9.11) and the discussion preceding it, we therefore have:

(10.10) PROPOSITION. Two right regular representations for a group G, obtained by different orderings of the list of elements of G, are equivalent.

EXERCISES

10.1. Find the right regular representation for Z_5.

• *10.2.* Find the left and right regular representations for V_4 by the method at the beginning of this section (that is, *not* by using a suitably arranged group table), and observe that the two are identical.

10.3. Verify that if ζ is a primitive cube root of 1, then

$$U(g) = \begin{bmatrix} \zeta & 0 & 0 \\ 0 & \zeta^2 & 0 \\ 0 & 0 & 1 \end{bmatrix}$$

gives a representation for $Z_3 = \langle g \rangle$. Then find a matrix C with complex entries such that $CT(g) = U(g)C$, where T is the right regular representation found in the text. Then, without performing any explicit matrix multiplications, show that $CT(g^2) = U(g^2)C$ and $CT(1) = U(1)C$. What can you conclude about T and U?

10.4. Carry out the computation for (10.7).

• *10.5.* Consider a family of matrices of the form

$$\begin{bmatrix} \bar{0} & A \\ B & \bar{0} \end{bmatrix},$$

where A and B are n-by-n submatrices, and $\bar{0}$ denotes an n-by-n matrix of zeros. Show that the product of any two such matrices has the form

$$\begin{bmatrix} C & \bar{0} \\ \bar{0} & D \end{bmatrix}$$

where C and D are n-by-n submatrices.

10.6. For the representation T given by (10.6) and (10.7), find the remaining four matrices for T, and verify that T preserves the defining relations for D_3. (*Hint:* Use exercises 9.4 and 10.5 to simplify the computations.)

10.7. Let $D_3 = \{1, r, r^2, c, rc, r^2c\}$ and write down a group table with the columns corresponding to the elements in the order given and the rows corresponding to their respective inverses (of course four of the elements are their own inverses). Verify that the right regular representation as computed in the text and exercise 10.6 may be read off the group table as described.

10.8. Let $G = \langle g : g^4 = 1 \rangle$ and form right regular representations T and U by taking the elements of G in the orders $\{1, g, g^2, g^3\}$ and $\{1, g, g^3, g^2\}$, respectively. Prove that T and U are equivalent by finding a nonsingular matrix A such that $AT(g) = U(g)A$.

Section 11

Irreducible Representations

In section 9 we found a representation V of degree 3 for D_4 such that *every* matrix $V(g)$ for $g \in D_4$ had the form

$$V(g) = \left[\begin{array}{cc|c} \multicolumn{2}{c|}{\multirow{2}{*}{$T(g)$}} & 0 \\ & & 0 \\ \hline 0 & 0 & U(g) \end{array}\right]$$

where T and U were representations of D_4 of degree 2 and 1, respectively. Now consider the vector subspaces

$$S_1 = \{(x, y, 0) : x, y \in \mathbf{R}\},$$
$$S_2 = \{(0, 0, z) : z \in \mathbf{R}\}$$

of $\mathbf{R}^3$, which are simply the XY-plane and the Z-axis. From linear algebra we know that $\mathbf{R}^3$ is the *direct sum* of S_1 and S_2, that is, that every vector in $\mathbf{R}^3$ can be written in exactly one way as a sum $s_1 + s_2$ with $s_1 \in S_1$ and $s_2 \in S_2$. For the direct sum we write

$$\mathbf{R}^3 = S_1 \oplus S_2.$$

If $v \in S_1$, then $vV(g) \in S_1$ also; likewise if $w \in S_2$, then $wV(g) \in S_2$ as well, for any $g \in D_4$. We say that the subspaces S_1 and S_2 are *invariant under* V.

In section 9 we noted informally that V decomposes into two smaller representations. Let us now make this concept explicit.

(11.1) DEFINITION. Let T be a representation of G. We call T *decomposable* if there are representations A and B of G such that T is equivalent to the representation

$$T^*(g) = \begin{bmatrix} A(g) & \bar{0} \\ \bar{0} & B(g) \end{bmatrix}, \qquad g \in G,$$

where the $\bar{0}$'s denote rectangular arrays of zeros of the required size for T^* to be square, as in (9.3). If T is not equivalent to any such representation, then T is called *indecomposable.*

As we remarked in section 9, if T is equivalent to a representation of the form

$$\begin{bmatrix} A(g) & \bar{0} \\ \bar{0} & B(g) \end{bmatrix}, \tag{11.2}$$

it is automatic that A and B are representations of G. Thus to show that a given T is decomposable, we need to prove only that it is equivalent to a representation T^* having the form (11.2) for all $g \in G$.

Note that if a representation T of G has the form (11.2) for every $g \in G$ (or, in view of exercise 9.4, if T has the form (11.2) for each element in a set of generators for G), then it is obviously decomposable. On the other hand, to show that T is indecomposable, we have to show that *T is not equivalent to any T^** having the form (11.2).

Recall from section 9 that for a representation T of a group G, if $T(g) \in GL(n, \mathbf{R})$ for all $g \in G$, we say that Euclidean n-dimensional space $\mathbf{R}^n$ *affords* T since each $T(g)$ specifies a linear transformation on $\mathbf{R}^n$. If some of the $T(g)$ have complex entries, then the space $\mathbf{C}^n$ of n-dimensional vectors with complex entries *affords* T. Of course, even if all $T(g)$ have only real entries, it is still possible to think of the $T(g)$ as linear transformations of $\mathbf{C}^n$, and we shall have occasion to do so in this section. Hence, in the discussion that follows, we must allow for the fact that either $\mathbf{R}^n$ or $\mathbf{C}^n$ may afford a representation T.

Now in (11.1), suppose that A has degree n and B has degree m. Then T has degree $n+m$. Let $\{e_1, e_2, \ldots, e_{n+m}\}$ be the natural basis for $\mathbf{R}^{n+m}$ (or for $\mathbf{C}^{n+m}$), let S_1 be the subspace spanned by $\{e_1, \ldots, e_n\}$, and let S_2 be the subspace spanned by $\{e_{n+1}, \ldots, e_{n+m}\}$. Then

$$\mathbf{R}^{n+m} = S_1 \oplus S_2 \qquad (\text{or } \mathbf{C}^{n+m} = S_1 \oplus S_2),$$

and each $T(g)$ carries vectors of S_1 to vectors of S_1 and vectors of S_2 to vectors of S_2.

We can abstract part of this observation as follows:

(11.3) DEFINITION. Let T be a representation of G having degree n, and let S be a subspace of $\mathbf{R}^n$ or of $\mathbf{C}^n$ with $1 \leq \dim S < n$. If $T(g)$ carries vectors of S to vectors of S for all $g \in G$, then S is called a *G-subspace* of $\mathbf{R}^n$ or of $\mathbf{C}^n$.

If in (11.3) we allowed dim S to be n, then S would be all of $\mathbf{R}^n$ or $\mathbf{C}^n$, which is trivially a G-subspace of itself.

Let us now connect the idea of a G-subspace with that of a decomposable representation:

(11.4) THEOREM. A representation T of G is decomposable if and only if the space $\mathbf{R}^n$ (or $\mathbf{C}^n$) affording T has G-subspaces S_1 and S_2 such that $\mathbf{R}^n = S_1 \oplus S_2$ (or $\mathbf{C}^n = S_1 \oplus S_2$).

PROOF. We give the proof for $\mathbf{R}^n$; the same argument holds for $\mathbf{C}^n$. If T is decomposable, then there is a change of basis that brings T to the form T^* of (11.2). Then S_1 and S_2 can be formed as in the discussion preceding (11.3), in terms of the new basis for $\mathbf{R}^n$. Conversely, if G-subspaces S_1 and S_2 exist with $\mathbf{R}^n = S_1 \oplus S_2$, then by a change of basis T can be brought into the form (11.2) as follows. Let $\{x_1, \ldots, x_k\}$ and $\{x_{k+1}, \ldots, x_n\}$ be $\mathbf{R}^n$-bases for S_1 and S_2, respectively. Then the matrix C whose rows are the vectors $x_1, \ldots, x_k, x_{k+1}, \ldots, x_n$ represents the change from the natural basis to the basis $\{x_1, \ldots, x_n\}$ for $\mathbf{R}^n$, and by hypothesis $CT^*(g) = T(g)C$ for every $g \in G$, where T^* has the form (11.2). This completes the proof.

Note that if we have a matrix C such that

$$CT^*(g) = T(g)C \quad \text{for all} \quad g \in G,$$

then since C is nonsingular, C^{-1} exists and $T^*(g) = C^{-1}T(g)C$, whence

$$C^{-1}T(g) = T^*(g)C^{-1} \quad \text{for all} \quad g \in G.$$

This shows that equivalence of representations is *symmetric*, that is, T is equivalent to T^* if and only if T^* is equivalent to T. Related concepts are raised in exercise 11.2.

It is important to realize that if $\mathbf{R}^n$ has a G-subspace S_1 for a representation T of G, then $\mathbf{R}^n$ will always have a *subspace* S_2 such that $\mathbf{R}^n = S_1 \oplus S_2$ as a direct sum of vector spaces, but S_2 *may fail to be a G-subspace* of $\mathbf{R}^n$. (Of course a similar statement can be made about $\mathbf{C}^n$.) For example, consider the regular representation T of V_4 (found in exercise 10.2); if $V_4 = \langle x, y : x^2 = y^2 = 1, yx = xy \rangle$, then T is given by

$$(11.5) \qquad T(x) = \begin{bmatrix} 0 & 1 & 0 & 0 \\ 1 & 0 & 0 & 0 \\ 0 & 0 & 0 & 1 \\ 0 & 0 & 1 & 0 \end{bmatrix} \qquad T(y) = \begin{bmatrix} 0 & 0 & 1 & 0 \\ 0 & 0 & 0 & 1 \\ 1 & 0 & 0 & 0 \\ 0 & 1 & 0 & 0 \end{bmatrix}.$$

Now let $S_1 = \{(a, a, a, a) : a \in \mathbf{R}\}$; then $T(x)$ and $T(y)$ merely carry each point of S_1 to itself, and since S_1 is obviously a subspace of dimension 1, it is a V_4-subspace. If we take $S_2 = \{(0, b, c, d) : b, c, d \in R\}$, then $\mathbf{R}^4 = S_1 \oplus S_2$, but S_2 is *not* a V_4-subspace—observe that $(0, b, c, d)T(x) = (b, 0, d, c)$, which is not in S_2 when $b \neq 0$. In this example, it happens that we can make a more fortuitous choice of S_2 so that T turns out to be decomposable, as we shall show later in this section.

To give an example of a representation that is indecomposable but that has a (proper) G-subspace, we let $G = \langle x \rangle$ be an infinite cyclic group, and consider the representation given by

$$T(x) = \begin{bmatrix} 1 & 1 \\ 0 & 1 \end{bmatrix}.$$

We recall from the end of section 8 that $\langle T(x) \rangle \cong \langle x \rangle$. Now for any integer k (positive, negative, or zero),

$$T(x^k) = \begin{bmatrix} 1 & 1 \\ 0 & 1 \end{bmatrix}^k = \begin{bmatrix} 1 & k \\ 0 & 1 \end{bmatrix}.$$

Let $S_1 = \{(0, b) : b \in \mathbf{R}\}$; then for any vector in S_1, $(0, b)T(x^k) = (0, b)$, so S_1 is a one-dimensional G-subspace of $\mathbf{R}^2$. Now if T is decomposable, then there must exist scalars λ, μ and a nonsingular matrix

$$A = \begin{bmatrix} a & b \\ c & d \end{bmatrix}$$

such that

$$\begin{bmatrix} a & b \\ c & d \end{bmatrix} \begin{bmatrix} 1 & 1 \\ 0 & 1 \end{bmatrix} = \begin{bmatrix} \lambda & 0 \\ 0 & \mu \end{bmatrix} \begin{bmatrix} a & b \\ c & d \end{bmatrix},$$

that is,

$$(11.6) \qquad \begin{bmatrix} a & a+b \\ c & c+d \end{bmatrix} = \begin{bmatrix} \lambda a & \lambda b \\ \mu c & \mu d \end{bmatrix}$$

and $ad - bc \neq 0$. If such scalars and A exist, we shall have shown that T is equivalent to a representation given by

$$T^*(x) = \begin{bmatrix} \lambda & 0 \\ 0 & \mu \end{bmatrix}.$$

Equating entries in (11.6), we have $\lambda a = a$, so either $\lambda = 1$ *or* $a = 0$. If $\lambda = 1$, then $a + b = \lambda b = b$, so $a = 0$. On the other hand, if $a = 0$, then $ad - bc \neq 0$ gives $bc \neq 0$, which means that $b \neq 0$ and $c \neq 0$. But we also have $\lambda b = a + b = b$ (using $a = 0$), so $\lambda = 1$ (since $b \neq 0$).

Thus $\lambda = 1$ *and* $a = 0$ (since each of these conditions implies the other). We thus have

$$\begin{bmatrix} 0 & b \\ c & c+d \end{bmatrix} = \begin{bmatrix} 0 & b \\ \mu c & \mu d \end{bmatrix}$$

and $bc \neq 0$. Again equating entries in (11.6), we see that $\mu c = c$ yields $\mu = 1$, and then $c + d = \mu d = d$ gives $c = 0$, which contradicts $bc \neq 0$ as found above. Therefore, T is indecomposable even though it has a proper G-subspace.

This example leads us to formulate the following:

(11.7) DEFINITION. Let T be a representation of G; T is *reducible* if there exist representations A and B of G such that T is equivalent to the representation

$$T^*(g) = \begin{bmatrix} A(g) & Q(g) \\ \bar{0} & B(g) \end{bmatrix}, \qquad g \in G, \tag{11.8}$$

where $\bar{0}$ is a submatrix of zeros and $Q(g)$ is a matrix (in general not square) depending upon g. If T is not equivalent to such a representation, then T is *irreducible.*

Note that in this definition $Q(g)$ has as many rows as $A(g)$ and as many columns as $B(g)$. Even if Q is square it is never a representation of G. For an illustration, let A, B, and Q be n-by-n and let g, $h \in G$; then

$$\begin{bmatrix} A(g) & Q(g) \\ \bar{0} & B(g) \end{bmatrix} \begin{bmatrix} A(h) & Q(h) \\ \bar{0} & B(h) \end{bmatrix} = \begin{bmatrix} A(g)A(h) & A(g)Q(h) + Q(g)B(h) \\ \bar{0} & B(g)B(h) \end{bmatrix},$$

which shows the homomorphism in the upper left and lower right corners, and a sum involving A and B as well as Q in the upper right corner of the product.

The irreducible representations of a group G are the building blocks of all representations of G and will be a main focus of our study.

Directly from the relevant definitions we have:

(11.9) PROPOSITION. If a representation T is decomposable, then T is reducible. Equivalently, if T is irreducible, then T is indecomposable.

The example of infinite cyclic G above shows that a representation may be reducible and indecomposable. My use of an infinite group for this example is not coincidental; in fact, the situation for finite groups is happier, as was shown by Maschke in 1898. In order to formulate Maschke's result, we must consider a basic idea:

(11.10) DEFINITION. A representation T of G is *completely reducible* if whenever T has a G-subspace S_1, T also has a G-subspace S_2 such that $\mathbf{R}^n = S_1 \oplus S_2$ (or $\mathbf{C}^n = S_1 \oplus S_2$), where n is the degree of T.

In terms of matrices, a representation T is completely reducible if whenever it can be put (by means of equivalence) into the form (11.8), it can also be put into the form (11.2). But bear in mind that putting T into one of these forms means finding a *single* change-of-basis matrix that puts *all* $T(g)$ into the proper form.

From (11.7) and (11.10) we have the linguistically amusing though mathematically trivial observation that *every irreducible representation is completely reducible.* (The point is that a representation that is completely reducible may have that property by virtue of having *no* proper G-subspace.) Of vastly more consequence is:

(11.11) MASCHKE'S THEOREM (special case). If G is a finite group and T is a representation of G afforded by $\mathbf{R}^n$ (or $\mathbf{C}^n$), then T is completely reducible.

Since we are concerned here only with representations using matrices from $GL(n, \mathbf{R})$ and $GL(n, \mathbf{C})$, I have stated this theorem in a restricted form. For the general form as well as a proof, see Curtis and Reiner [**4**], p. 41.

To study the implications of Maschke's Theorem, we let T be a reducible representation of a finite group G. Then T is equivalent to a representation T^* of the form (11.8), and so to a representation T^{**} of the form (11.2). Now the A and B in (11.2) are also completely reducible by Maschke's Theorem. Thus either A is irreducible or T^{**} is equivalent to a representation T^{***} of the form

$$(11.12) \qquad T^{***}(g) = \begin{bmatrix} A_1(g) & \bar{0} & \bar{0} \\ \bar{0} & A_2(g) & \bar{0} \\ \bar{0} & \bar{0} & B(g) \end{bmatrix}, \qquad g \in G,$$

and a similar comment applies to B. We may repeat this process a finite number of times (which number obviously cannot exceed the degree of T) to obtain a representation U equivalent to T and having the form

$$(11.13) \qquad U(g) = \begin{bmatrix} A_1(g) & \bar{0} & \dots & \bar{0} \\ \bar{0} & A_2(g) & \dots & \bar{0} \\ & \dots & & \dots \\ \bar{0} & \bar{0} & \dots & A_k(g) \end{bmatrix}, \qquad g \in G,$$

in which $A_1, \dots, A_k$ are all irreducible. Here we have suppressed B of (11.12) and any B_1, B_2, etc., into which it is decomposed, and simply listed as A_j $(1 \leq j \leq k)$ the irreducible representations at which we arrive. Thus we have:

(11.14) THEOREM. Every (real or complex) representation T of a finite group G may be decomposed into *irreducible components*, that is, there are irreducible representations $A_1, \dots, A_k$ of G such that T is equivalent to a representation U having the form (11.13).

PROOF. This result follows from the discussion above together with the observation that equivalence of matrices is transitive, a detail relegated to exercise 11.2.

As pointed out in connection with (9.3), the converse holds:

(11.15) PROPOSITION. If $A_1, \dots, A_k$ are irreducible representations of G and if U is formed as at (11.13), then U is a representation of G.

We can illustrate Maschke's Theorem for the regular representation of V_4, as given at (11.5). From exercise 9.2 we have the following as the complete list of one-dimensional representations of V_4:

$$(11.16) \qquad \begin{cases} T_1(x) = [\ \ 1], & T_1(y) = [\ \ 1], \\ T_2(x) = [\ \ 1], & T_2(y) = [-1], \\ T_3(x) = [-1], & T_3(y) = [\ \ 1], \\ T_4(x) = [-1], & T_4(y) = [-1]. \end{cases}$$

Now let

$$C = \begin{bmatrix} 1 & 1 & 1 & 1 \\ 1 & 1 & -1 & -1 \\ -1 & 1 & -1 & 1 \\ -1 & 1 & 1 & -1 \end{bmatrix};$$

Since $\det(C) = -16$, C is nonsingular, as required for equivalence, and by performing the matrix products, we can obtain

$$CT(x) = \begin{bmatrix} 1 & 0 & 0 & 0 \\ 0 & 1 & 0 & 0 \\ 0 & 0 & -1 & 0 \\ 0 & 0 & 0 & -1 \end{bmatrix} C \quad \text{and} \quad CT(y) = \begin{bmatrix} 1 & 0 & 0 & 0 \\ 0 & -1 & 0 & 0 \\ 0 & 0 & 1 & 0 \\ 0 & 0 & 0 & -1 \end{bmatrix} C.$$

Hence the regular representation T is equivalent to a representation whose irreducible components are precisely the four one-dimensional representations of V_4.

Since a representation of degree 1 is obviously irreducible, it will be helpful at this point to consider an example of an irreducible representation of degree greater than 1. At (9.4) and (9.5) we found a representation of the group G of rigid symmetries of the cube, given by

$$T(r) = \begin{bmatrix} 0 & -1 & 0 \\ 1 & 0 & 0 \\ 0 & 0 & 1 \end{bmatrix} \qquad T(d) = \begin{bmatrix} 0 & -1 & 0 \\ 0 & 0 & -1 \\ 1 & 0 & 0 \end{bmatrix}.$$

Now by (11.11), if T is reducible, it must be equivalent to a representation having the form

$$T^*(g) = \begin{bmatrix} U(g) & \bar{0} \\ \bar{0} & V(g) \end{bmatrix}$$

where U has degree 2 and V has degree 1. (By exercise 11.5 we need not consider the case in which U has degree 1 and V has degree 2.) Suppose that T is equivalent to such a T^*; let $C = (c_{ij})$ be a nonsingular matrix, $V(r) = [\lambda]$ and $V(d) = [\mu]$, and $CT(g) = T^*(g)C$ for $g \in G$. From the *third row* of the products

$$C \begin{bmatrix} 0 & -1 & 0 \\ 1 & 0 & 0 \\ 0 & 0 & 1 \end{bmatrix} = \begin{bmatrix} \multicolumn{2}{c}{U(r)} & \begin{matrix} 0 \\ 0 \end{matrix} \\ 0 & 0 & \lambda \end{bmatrix} C$$

we obtain

(11.17) $\qquad c_{32} = \lambda c_{31}, \qquad -c_{31} = \lambda c_{32}, \qquad c_{33} = \lambda c_{33},$

and from the *third row* of the products

$$C \begin{bmatrix} 0 & -1 & 0 \\ 0 & 0 & -1 \\ 1 & 0 & 0 \end{bmatrix} = \begin{bmatrix} \multicolumn{2}{c}{U(d)} & \begin{matrix} 0 \\ 0 \end{matrix} \\ 0 & 0 & \mu \end{bmatrix} C$$

we obtain

$$(11.18) \qquad c_{33}=\mu c_{31}, \qquad -c_{31}=\mu c_{32}, \qquad -c_{32}=\mu c_{33}.$$

From (11.17) we conclude that $\lambda=1$ *or* $c_{33}=0$. But if $c_{33}=0$, then by (11.18), $c_{31}=c_{32}=0$, which gives C an entire row of zeros and makes C singular, contrary to our assumption. Hence $\lambda=1$. But then (11.17) gives $c_{31}=c_{32}=0$, as the reader should verify, and by (11.18) we get $c_{33}=0$ also, so once again we have contradicted the assumption that C is nonsingular. Since both attempts lead to a contradiction, no such C can exist, and T must be irreducible.

In this example we made no assumption as to whether the c_{ij} were real or complex. To conclude this section let us consider an example of a representation that is irreducible over the real but reducible over the complex numbers. Let $G=\langle g\rangle$ be cyclic of order 3, and take

$$U(g)=\begin{bmatrix}0 & -1\\ 1 & -1\end{bmatrix};$$

then $U(g)$ determines a representation U of G. If U is reducible over the real numbers, there must exist a nonsingular matrix A and real scalars λ and μ with

$$(11.19) \qquad AU(g)=\begin{bmatrix}\lambda & 0\\ 0 & \mu\end{bmatrix}A.$$

Let $|A|$ denote the determinant of A; since the determinant of a product is equal to the product of the determinants, we have $|A|\cdot 1=\lambda\mu|A|$ since $U(g)$ has determinant 1. Since $|A|\neq 0$, we have $\lambda\mu=1$.

Recall from linear algebra that the *trace* of a square matrix is the sum of the elements on its main diagonal. In general, the trace of a product is *not* equal to the product of the traces; however, we shall see in section 13 that if there is a nonsingular matrix C such that $B=C^{-1}AC$, then A and B have the same trace. Thus if (11.19) holds, then $U(g)$ must have trace $\lambda+\mu$, that is, $\lambda+\mu=-1$. But then

$$\lambda\mu=\lambda(-1-\lambda)=1,$$

whence

$$\lambda^2+\lambda+1=0,$$

which equation has no real root. Therefore, U is irreducible over the reals. The fact that U is reducible over the complex numbers is left as exercise 11.7.

EXERCISES

11.1. Let $\{e_1, e_2, e_3, e_4\}$ be the natural basis for $\mathbf{R}^4$, and let

$$x_1 = 4e_1 + e_2 + e_3 + e_4,$$
$$x_2 = -3e_1 + 3e_2,$$
$$x_3 = -3e_1 + 3e_3,$$
$$x_4 = -3e_1 + 3e_4.$$

Find the matrix C for the change of basis from $\{e_i\}$ to $\{x_i\}$ and verify that the representation T of V_4 given by (11.5) is equivalent to the representation T^* determined by

$$T^*(x) = \begin{bmatrix} 1 & 1 & 0 & 0 \\ 0 & -1 & 0 & 0 \\ 0 & -1 & 0 & 1 \\ 0 & -1 & 1 & 0 \end{bmatrix}, \qquad T^*(y) = \begin{bmatrix} 1 & 0 & 1 & 0 \\ 0 & 0 & -1 & 1 \\ 0 & 0 & -1 & 0 \\ 0 & 1 & -1 & 0 \end{bmatrix}.$$

Then verify that $S_1 = \{rx_1 : r \in \mathbf{R}\}$ is a V_4-subspace of $\mathbf{R}^4$ with respect to T^*. Finally, let S_2 be the span of $\{x_2, x_3, x_4\}$, and show that $\mathbf{R}^4 = S_1 \oplus S_2$ but that S_2 is *not* a V_4-subspace.

11.2. Let T, T^*, and T^{**} be representations of G such that T is equivalent to T^* and T^* is equivalent to T^{**}. Prove that T is equivalent to T^{**}. (This result is the transitivity of equivalence of representations. Clearly, equivalence is reflexive and we showed, following the proof of (11.4), that equivalence is symmetric; hence equivalence of representations is an *equivalence relation.*)

11.3. Let ζ be a primitive k^{th} root of 1 (see exercise 8.4), let $G = \langle x : x^{2k} = 1 \rangle$, and let U be a representation of G given by

$$U(x) = \begin{bmatrix} 0 & \zeta \\ 1 & 0 \end{bmatrix}.$$

Verify that U is indeed a representation of G, and find its irreducible components. Then check that U is *not* a representation of a cyclic group of order k.

11.4. In section 9 we found a representation of

$$V_4 = \langle x, y : x^2 = y^2 = 1, yx = xy \rangle$$

given by

$$T(x) = \begin{bmatrix} 0 & 1 \\ 1 & 0 \end{bmatrix}, \qquad T(y) = \begin{bmatrix} -1 & 0 \\ 0 & -1 \end{bmatrix}.$$

Find the irreducible components of T.

11.5. Let A and B be representations of G, and for each $g \in G$, let

$$T(g) = \begin{bmatrix} A(g) & \bar{0} \\ \bar{0} & B(g) \end{bmatrix} \quad \text{and} \quad T^*(g) = \begin{bmatrix} B(g) & \bar{0} \\ \bar{0} & A(g) \end{bmatrix}.$$

Prove that T is equivalent to T^*.

11.6. Derive (11.17) and (11.18).

11.7. Find a nonsingular matrix A and *complex* scalars λ, μ such that (11.19) holds. (*Hint:* Use the equation $\lambda^2 + \lambda + 1 = 0$ derived in the text.)

11.8. Let S_1 be the subspace of $\mathbf{R}^4$ defined immediately following (11.5). By Maschke's Theorem, we know that there is a V_4-subspace U of $\mathbf{R}^4$ such that $\mathbf{R}^4 = S_1 \oplus U$. Show that U may be taken explicitly to be the subspace spanned by the vectors

$$(1, 1, -1, -1), (1, -1, 1, -1), \quad \text{and} \quad (1, -1, -1, 1).$$

11.9. Use determinants and traces to give an alternative proof of the fact that the representation of $Z_\infty = \langle x \rangle$ given by

$$T(x) = \begin{bmatrix} 1 & 1 \\ 0 & 1 \end{bmatrix}$$

is indecomposable.

Section 12
Representations of Abelian Groups

We have already encountered a number of representations of abelian groups of various orders (including the infinite). In this section we find all *one-dimensional* representations of *finite* abelian groups; in a later section we shall see how these "little" representations fit into the problem of determining *all* representations of finite abelian groups.

First, let $G=\langle x\rangle$ be a cyclic group of order k, and recall that a complex number ζ is a *primitive* k^{th} *root of* 1 if $\zeta^k=1$ and if $\zeta^m\neq 1$ whenever $1\leq m<k$. Now to specify a representation T of G, we need only give $T(x)$ and verify that $T(x)^k=I_n$, where n is the degree of T. Since we are presently interested in the case $n=1$, we merely require that $T(x)^k=[1]$.

Now if $1\leq j\leq k$ and $T_j(x)=[\zeta^j]$, then

$$T_j(x)^k=[\zeta^j]^k=[\zeta^{jk}],$$

which is [1], as required. Conversely, if $T(x)=[\phi]$ for some complex number ϕ, then $T(x)^k=[1]$ requires $\phi^k=1$, and hence ϕ must be equal to ζ^j for some j between 1 and k.

(12.1) Proposition. If G is cyclic of order k, and if ζ is a primitive k^{th} root of 1, then for $j=1,\ldots,k$, $T_j(x)=[\zeta^j]$ is a complete list of the one-dimensional representations of G.

This result has been proved in the discussion preceding it, except for the detail of showing that if $j\neq m$, then T_j is not equivalent to T_m (see exercise 12.1); hence there are exactly k one-dimensional representations of a cyclic group of order k.

Before proceeding to the representations of noncyclic abelian groups, we need to consider some additional material from abstract group theory:

(12.2) Definition. Let H and K be normal subgroups of G with $H\cap K=\{1\}$ and $G=HK$. Then G is called the *direct product* of H and K; we write $G=H\times K$.

The direct product of groups (or subgroups, as above) is analogous to the direct sum of two vector subspaces. Two equivalent formulations are:

(12.3) PROPOSITION. Let H and K be subgroups of G (not assumed normal). Then G is the direct product of H and K if and only if $G = HK$, $H \cap K = \{1\}$, and $hk = kh$ for each $h \in H$ and each $k \in K$.

(12.4) PROPOSITION. Let H and K be *normal* subgroups of G. Then G is the direct product of H and K if and only if each element of G can be written in exactly one way as a product hk with $h \in H$ and $k \in K$.

We met direct products at the end of section 6; the discussion of a group G of order 35 amounted to showing that G was the direct product of a subgroup of order 5 and a subgroup of order 7. Exercise 6.1 asked you to find that a group of order 15 must be the direct product of a subgroup of order 3 and one of order 5. In both instances, the resulting group was cyclic; later in this section we prove a result linking direct products of subgroups having relatively prime order with cyclic groups.

The procedure of taking a group and writing it as a direct product of two subgroups can be reversed to take two groups and form a new group that is their direct product, as follows:

(12.5) PROPOSITION. Let H and K be groups, let

$$G = \{(h, k) : h \in H, k \in K\}$$

with the operation $(h, k)(h', k') = (hh', kk')$, and let $H^* = \{(h,1) : h \in H\}$ and $K^* = \{(1,k) : k \in K\}$. Then G is a group, H^* and K^* are subgroups of G, $H \cong H^*$, $K \cong K^*$, and $G = H^* \times K^*$.

Since (12.5) forms a new group as the direct product of two given groups, its G is sometimes called the *external direct product* of H and K, whereas the G in (12.2) is called the *internal direct product* of H and K. The two views are related by the observation that G in (12.5) is the internal direct product of H^* and K^*, and G in (12.2) is isomorphic to the external direct product of H and K (which is easily proved using (12.4)). We make the latter part of this observation explicit as follows:

(12.6) PROPOSITION. Let $G = H \times K$; form

$$G^* = \{(h, k) : h \in H, k \in K\},$$

define the operation on G^* as

$$(h, k)(h', k') = (hh', kk'),$$

and let H^* and K^* be as in (12.5). Then $H^* \cong H$, $K^* \cong K$, and $G^* \cong G$.

Now suppose that $G = H \times K$, T is a representation of H, and U is a representation of K. It would be convenient if we could take advantage of the decomposition of each $g \in G$ in a unique way as a product hk (as in (12.4), which is equivalent to the (h, k) of (12.6)) in order to define a representation V of G by

(12.7) $$V(g) = V(hk) = T(h)U(k),$$

for which purpose we would want to let T and U have the same degree. However, by (12.3), $hk = kh$ for any choice of $h \in H$ and $k \in K$, which means that

$$(h \cdot 1)(1 \cdot k) = (1 \cdot k)(h \cdot 1),$$

and thus for (12.7) to define a representation we would need to have

(12.8) $$T(h)U(k) = U(k)T(h)$$

for each $h \in H$ and $k \in K$. But (12.8) does not hold in general for matrices unless they are of degree 1. What we can say is:

(12.9) PROPOSITION. If $G = H \times K$ and if T and U are *one-dimensional* representations of H and K, respectively, then $V(hk) = T(h)U(k)$ is a representation of G.

For example, let G be the group of order 9 given by

$$G = \langle x, y : x^3 = y^3 = 1, yx = xy \rangle;$$

then $G = \langle x \rangle \times \langle y \rangle$. Both $\langle x \rangle$ and $\langle y \rangle$ are of order 3 and so the only one-dimensional representations available are (by (12.1)) those determined by

(12.10) $$\begin{cases} T_1(x) = [\zeta], & T_2(x) = [\zeta^2], \quad T_3(x) = [1]; \\ U_1(y) = [\zeta], & U_2(y) = [\zeta^2], \quad U_3(y) = [1], \end{cases}$$

where ζ is a primitive cube root of 1.

Conversely, if V is any one-dimensional representation of G, then the restriction of its domain to the subgroup $\langle x \rangle$ gives a representation of $\langle x \rangle$; hence $V(x) = T_j(x)$ for some j $(j = 1,2,3)$. Similarly, $V(y) = U_k(y)$ for some k. Thus V must agree with some

T_j on $\langle x \rangle$ and with some U_k on $\langle y \rangle$ and, since G consists of products of powers of x and y, V must be determined by the T_j and U_k. Hence *all* one-dimensional representations of G must be given by the nine possible combinations of T_j and U_k in (12.10). We denote such a combination by V_{jk} with $1 \leq j, k \leq 3$; specifically, take

(12.11) $$V_{jk}(x^r y^s) = T_j(x)^r U_k(y)^s$$

for $r = 0, 1, 2$ and $s = 0, 1, 2$. Now we note that if $V_{jk} = V_{mn}$ (as functions, that is, they take the same values on the same points), then

$$T_j(x) = V_{jk}(x) = V_{mn}(x) = T_m(x),$$

whence T_j is the same as T_m and $j = m$; similarly, $k = n$. Thus applying formula (12.11) to the representations in (12.10) yields nine *distinct* representations of G, and they constitute a complete list of the one-dimensional representations of G.

Happily, the argument used in this example is the prototype for finding all of the one-dimensional representations of a finite abelian group. I shall state the information we need about group structure without proof. First, to generalize our concept of the direct product:

(12.12) DEFINITION. Let $H_1, \ldots, H_n$ be normal subgroups of G and let each element $g \in G$ have a unique representation

$$g = h_1 h_2 \ldots h_n \quad \text{with} \quad h_1 \in H_1, \ldots, h_n \in H_n.$$

Then G is called the *direct product* of $H_1, \ldots, H_n$; we write

$$G = H_1 \times \ldots \times H_n.$$

Note that I have taken (12.4) rather than (12.2) as the model for this definition; the generalization of (12.2) to more than two factors is left as exercise 12.8. The analogues of (12.5) and (12.6) with n factors can be shown to hold.

Now, we are ready to consider a deep and powerful result:

(12.13) MAIN THEOREM ON FINITE ABELIAN GROUPS. If G is a finite abelian group, then G is isomorphic to a direct product of cyclic groups of prime-power order, that is, there exist primes p_j (not necessarily distinct) and positive integers a_j such that for $q_j = p_j^{a_j}$,

(12.14) $$G = Z_{q_1} \times \ldots \times Z_{q_n}.$$

(A group has prime-power order if its order is some positive power of a given prime number.)

A similar theorem holds for infinite abelian groups having only a finite number of generators; in this case the factors in (12.14) may be infinite cyclic as well as finite.

This theorem tells how we can determine all *abelian* groups of a given (finite) order. For example, if $|G| = p^2$, where p is a prime, then G must be isomorphic either to a cyclic group of order p^2 or to $Z_p \times Z_p$. In particular, the only abelian groups of order 4 are Z_4 and $Z_2 \times Z_2$, the latter being isomorphic to V_4. From exercise 7.16 we know that these two abelian groups are in fact the only groups of order 4. We can generalize this observation as follows:

(12.15) PROPOSITION. Let p be prime and $|G| = p^2$; then either G is cyclic or G is isomorphic to $Z_p \times Z_p$. Hence G is abelian.

PROOF. We observe first that the cyclic group of order p^2 is *not* isomorphic to $Z_p \times Z_p$ because isomorphisms preserve the orders of elements (exercise 5.11), but the cyclic group has elements of order p^2 and $Z_p \times Z_p$ does not. Obviously, both of these groups are abelian. Now suppose that G is not cyclic; by (6.2), $C(G)$ contains an element a different from 1. Hence the order of a must be p. Since $a \in C(G)$, $g^{-1}ag = a$ for all $g \in G$, and so $A = \langle a \rangle \trianglelefteq G$. If $b \notin A$, then b has order p, and the distinct cosets of A in G are A, Ab, $Ab^2, \ldots, Ab^{p-1}$, accounting for the p^2 elements. Let $B = \langle b \rangle$; then $G = AB$, $A \cap B = \{1\}$, and since $ab = ba$, $B \trianglelefteq G$. By (12.2), $G = A \times B$. This completes the proof.

Now let p and q be distinct primes and let $|G| = pq$. Then pq is not a prime power, and so the only possible way to write G in the form (12.14) is as $Z_p \times Z_q$. Now we know that there is a cyclic group of order pq because there is a cyclic group of every order, and a cyclic group is certainly abelian; hence it must happen that the cyclic group of order pq is isomorphic to $Z_p \times Z_q$. In fact, a more general result holds.

First, recall that two integers are *relatively prime* if their greatest common divisor is 1, that is, if they have no prime factors in common. Thus if p and q are distinct primes, then p and q are relatively prime, but a pair of composite integers such as 14 and 15 can also be relatively prime since they have no common divisor other than ± 1. Now we can state the general result:

(12.16) PROPOSITION. If m and n are relatively prime positive integers, then the cyclic group of order mn is isomorphic to $Z_m \times Z_n$.

PROOF. Let G be the external direct product of $\langle a : a^m = 1 \rangle$ and $\langle b : b^n = 1 \rangle$; then G may be written as

$$\langle a, b : a^m = b^n = 1, ba = ab \rangle.$$

Let $H = \langle x : x^{mn} = 1 \rangle$, and define $\sigma : H \to G$ by $\sigma(x) = ab$, that is, for $0 \le k < mn$, let $\sigma(x^k) = a^k b^k$. Then

$$\begin{aligned} \sigma(x^j x^k) &= \sigma(x^{j+k}) = a^{j+k} b^{j+k} \\ &= a^j b^j a^k b^k = \sigma(x^j)\sigma(x^k), \end{aligned}$$

so σ is a homomorphism. Let $x^k \in$ kernel (σ); then $a^k b^k = 1$, and hence both m and n must divide k. (See exercise 12.10.) Since m and n are relatively prime and $0 \le k < mn$, we must have $k = 0$, that is, $x^k = 1$; therefore, σ is 1:1. Now

$$|G| = |\langle a \rangle||\langle b \rangle| = mn = |H|,$$

and σ is 1:1, so σ is onto. This completes the proof.

We remarked at the end of exercise 6.1 that there are noncyclic groups of order $3 \cdot 7 = 21$; by (12.16) and (12.13), these groups must also be nonabelian.

Observe that if $H_1, \ldots, H_n$ are cyclic groups, then by exercise 12.9 together with the fact that cyclic groups are abelian, we have that $H_1 \times \ldots \times H_n$ is abelian. Hence the question for any given order is: How can $|G|$ be decomposed as a product of prime powers? The prime powers for such a decomposition are the q_j for (12.14), and thus a list of such decompositions of $|G|$ is in effect a list of the abelian groups of that order.

For example, if $|G| = 12$, then the only ways in which 12 can be factored as a product of prime powers are $4 \cdot 3$ and $2 \cdot 2 \cdot 3$, so any abelian group of order 12 must be isomorphic to $Z_4 \times Z_3$ or to $Z_2 \times Z_2 \times Z_3$. In particular, by (12.16) the cyclic group of order 12 is isomorphic to $Z_4 \times Z_3$. On the other hand, $Z_2 \times Z_6$, which is obviously an abelian group of order 12, is isomorphic to $Z_2 \times Z_2 \times Z_3$. (See exercise 12.12.)

The prime-power decompositions of 24 are $8 \cdot 3$, $4 \cdot 2 \cdot 3$, and $2 \cdot 2 \cdot 2 \cdot 3$; hence the three abelian abstract groups of order 24 are

$$Z_8 \times Z_3, \qquad Z_4 \times Z_2 \times Z_3 \quad \text{and} \quad Z_2 \times Z_2 \times Z_2 \times Z_3.$$

Further examples are considered in exercise 12.13.

By means of (12.9) and (12.13) we can easily determine all one-dimensional representations of a finite abelian group G. Let G be as in (12.14) and write $Z_{q_j} = \langle x_j \rangle$ for $j = 1, \ldots, n$. Let ζ_j be a primitive q_j^{th} root of 1. Then

$$T_j(x_j) = [\zeta_j]$$

gives a representation of $\langle x_j \rangle$ and for $g = x_1^{b_1} x_2^{b_2} \ldots x_n^{b_n} \in G$,

$$T(g) = [\zeta_1^{b_1} \zeta_2^{b_2} \ldots \zeta_n^{b_n}]$$

gives a representation of G. As we saw in (12.1), each $\langle x_j \rangle$ has exactly q_j one-dimensional representations, which we may denote by double subscripting T_j as

$$T_{j1}(x_j) = [\zeta_j], \qquad T_{j2}(x_j) = [\zeta_j^2], \qquad \ldots, \qquad T_{jq_j}(x) = [\zeta_j^q],$$

where of course T_{jq_j} is just the trivial representation. By the same reasoning as that used for $Z_3 \times Z_3$ we obtain $q_1 q_2 \ldots q_n$ distinct one-dimensional representations for G. But $|G| = q_1 q_2 \ldots q_n$. Hence we have proved:

(12.17) THEOREM. If G is a finite abelian group, then there are exactly $|G|$ one-dimensional representations of G over the complex numbers **C**

As we shall see later, the one-dimensional representations will be the *only* irreducible representations of a finite abelian group over **C**.

As a postscript to our discussion of abelian groups of order pq, where p and q are distinct primes, it will be illustrative to compare the results obtained by means of the proof of (12.17) for the one-dimensional representations of $Z_p \times Z_q$ with those obtained in our discussion of cyclic groups in (12.1). Let ζ be a primitive p^{th} root of 1 and η a primitive q^{th} root of 1. Then the one-dimensional representations of $Z_p \times Z_q$ are precisely those given by $T_{jk}(xy) = [\zeta^j \eta^k]$, where $Z_p = \langle x \rangle$ and $Z_q = \langle y \rangle$, and the subscripts run $0 \leq j \leq p-1$, $0 \leq k \leq q-1$. In particular, if we take $j = k = 1$, then the powers of $\zeta\eta$ give all pq of the distinct representations of degree 1 since $(\zeta\eta)^m = 1$ only when both p and q divide m. But this is the same as saying that $\zeta\eta$ is a primitive $(pq)^{\text{th}}$ root of 1, and we know that the one-dimensional representations of $Z_{(pq)}$ are given by

$$T_n(z) = [\zeta\eta]^n,$$

where $0 \leq n < pq$ and $Z_{(pq)} = \langle z \rangle$.

Thus, for example, if $p=2$ and $q=3$, we may take $\zeta=-1$ and $\eta=\frac{1}{2}(-1+i\sqrt{3})$; then $\zeta\eta=-\eta$ is indeed a primitive 6th root of 1, and the one-dimensional representations we get from the proof of (12.17) are the same as those we get from (12.1).

To conclude this section, let us show how the one-dimensional representations of *abelian* groups may be used to determine the one-dimensional representations of *any* finite group. Our first step is to show how any representation of a factor group G/N gives rise to a representation of the group G.

(12.18) PROPOSITION. Let $N \trianglelefteq G$ and let T be any representation of G/N with degree n. Define $\hat{T}$ on G by

$$\hat{T}(g)=T(Nx) \quad \text{if } g \text{ is in the coset } Nx.$$

Then $\hat{T}$ is a representation of G. Moreover, if U is a representation of G/N with U not equivalent to T, then $\hat{U}$ (formed as above) is not equivalent to $\hat{T}$.

For example, let

$$G=D_4=\langle r, c : r^4=c^2=1,\ cr=r^3c\rangle,$$

let $N=\langle r\rangle$, and consider the representation

$$T(N)=[1], \qquad T(Nc)=[-1]$$

on the factor group G/N, which is cyclic of order 2. Then

$$\hat{T}(r^i)=[1], \qquad \hat{T}(r^ic)=[-1]$$

for $i=0, 1, 2, 3$.

To apply this construction of a representation of G from one of G/N, we need the concept of the *derived group:*

(12.19) DEFINITION. The *derived group*, or *commutator subgroup* of a group G is the subgroup G' generated by all elements of the form $ghg^{-1}h^{-1}$ for $g, h \in G$.

Note that closure in G' is guaranteed by the phrase "generated by." The most basic facts about the derived group are given by the following:

(12.20) PROPOSITION. For any group G, the derived group G' is a normal subgroup. Moreover, if $N \trianglelefteq G$, then G/N is abelian if and only if $G' \leq N$.

A special case of (12.20) is the observation that G/G' is abelian; this fact establishes that there are at least $[G:G']$ one-dimensional representations of a finite group G (or even of an infinite group G with the property that G/G' is finite) by (12.18). Now we observe:

(12.21) PROPOSITION. Let T be any one-dimensional representation of a group G. Then T, with its domain restricted to the subgroup G', is the trivial representation of G'; moreover, T gives a one-dimensional representation of G/G' if we take $T^*(G'x) = T(x)$ for each $G'x \in G/G'$.

Finally, we have our main result:

(12.22) THEOREM. If G is finite, then G has exactly $[G:G']$ one-dimensional representations over the complex field $\mathbf{C}$.

We remark that this theorem holds under the weaker hypothesis that G/G' is a finite group. The proofs of (12.18), (12.20), (12.21), and (12.22) are straightforward and have been left as exercises.

EXERCISES

12.1. Let $G = \langle x : x^k = 1 \rangle$, let ζ be a primitive k^{th} root of 1, and let representations T and U of G be given by

$$T(x) = [\zeta^m], \qquad U(x) = [\zeta^n]$$

with $1 \le m, n \le k$. Prove that T and U are equivalent if and only if $m = n$.

12.2. Prove (12.3).

12.3. Give an example to show that if $H \trianglelefteq G$, $K \le G$, $H \cap K = \{1\}$, and $G = HK$, then G *need not* be $H \times K$.

12.4. Prove (12.4).

12.5. Give an example to show that if $H \trianglelefteq G$, $K \le G$, and each element of G can be written in one and only one way as hk with $h \in H$ and $k \in K$, then G *need not* be $H \times K$.

12.6. Prove (12.5).

12.7. Prove (12.6).

12.8. Let $H_1, \ldots, H_n$ be subgroups of G. We define the *join* of these subgroups to be the intersection of all subgroups of G that contain all of $H_1, \ldots, H_n$. (The group G is such a subgroup, so the intersection is not empty. By exercise 3.1, an intersection of subgroups is again a subgroup, so the join is a subgroup of G.) Now let the H_j be normal subgroups of G. Prove that G is the direct product of the H_j if and only if

(a) G is the join of $H_1, \ldots, H_n$;

and

(b) for each j, the intersection of H_j with the join of all H_k *except* H_j is $\{1\}$.

12.9. Let $H_1, \ldots, H_n$ be subgroups of G. Prove that $G = H_1 \times \ldots \times H_n$ if and only if

(a) each element $g \in G$ has a unique expression as a product $h_1 h_2 \ldots h_n$ with $h_1 \in H_1, \ldots, h_n \in H_n$;

and

(b) if $h_j \in H_j$, $h_k \in H_k$, and $j \neq k$, then $h_j h_k = h_k h_j$.

12.10. Let $a, b \in G$ with orders m, n respectively, let m and n be relatively prime, and assume that $ba = ab$. Prove that if $(ab)^k = 1$, then both m and n divide k. (*Hint:* Prove first that if $a^k = 1$, then the order of a divides k. To do so, let m be the order of a, and use the fact that there exist a quotient and remainder q and r with $0 \leq r < m$ such that $k = qm + r$. Then let $(ab)^k = 1$, show that $a^k \in \langle b \rangle$ and so $a^k = 1$; similarly, $b^k = 1$.)

12.11. Show that the conclusion of exercise 12.10 *need not* hold if the assumption that m and n are relatively prime is dropped. (*Hint:* An example can be found using Z_4.)

12.12. Verify that $Z_2 \times Z_6 \cong Z_2 \times Z_2 \times Z_3$.

12.13. Use (12.13) to find all abelian groups of orders 16, 18, and 36. (The number of groups is 5, 2, and 4, respectively.)

12.14. Find the derived group of D_4 and of Q_2.

12.15. Prove (12.20). Note that for the first part of the proposition, it suffices to prove that the conjugate of a generator $ghg^{-1}h^{-1}$ by an arbitrary element x is in G' in order to show that G' is normal.

12.16. Prove (12.18).

12.17. Prove (12.21). Note that it is necessary to show that T^* is well defined, that is, if $G'x = G'y$, then $T^*(G'x) = T^*(G'y)$.

12.18. Use (12.18) and (12.21) to prove (12.22). (*Hint:* It will be necessary to establish the remark in the text that there are at least $[G:G']$ one-dimensional representations of G. Then (12.21) may be used to show that equality holds.)

Part Three
CHARACTERS

Section 13
Group Characters

Recall from the end of section 11 that the *trace* of a square matrix is the sum of the elements on the main diagonal; we denote the trace of A by $\mathrm{tr}(A)$. Now let T be a representation of G, and for each $g \in G$, let

$$\chi(g) = \mathrm{tr}(T(g)). \tag{13.1}$$

We call χ a *character* of G, in particular, the *character associated with* T. Thus a character is a function $\chi: G \to \mathbf{R}$ or $\chi: G \to \mathbf{C}$, depending upon whether T is afforded by $\mathbf{R}^n$ or $\mathbf{C}^n$.

For example, let G be finite and let T be the (left or right) regular representation of G. Then from (10.9) we see that $T(1)$ has ones down its main diagonal, and $T(g)$ has all zeros down its main diagonal if $g \neq 1$. Thus

$$\begin{cases} \chi(1) = |G|, \\ \chi(g) = 0 \quad \text{if} \quad g \neq 1 \end{cases}$$

is the character associated with the regular representation of *any* finite G.

Clearly, if T is any representation of a group G with degree $(T) = 1$, then T is related to its associated character χ by

$$T(g) = [\chi(g)], \qquad g \in G.$$

In order to study characters, we must recall some ideas about the trace from linear algebra. In general, $\mathrm{tr}(AB)$ is *not* equal to

$\text{tr}(A)\text{tr}(B)$. However, we do have:

(13.2) PROPOSITION. If A and B are n-by-n matrices, then $\text{tr}(AB) = \text{tr}(BA)$.

PROOF. Let $A = (a_{ij})$ and $B = (b_{ij})$ with $1 \le i, j \le n$. Then

$$\begin{aligned} \text{tr}(AB) &= \text{tr}\left(\sum_{k=1}^{n} a_{ik}b_{kj}\right) = \sum_{i=1}^{n}\sum_{k=1}^{n} a_{ik}b_{ki} \\ &= \sum_{k=1}^{n}\sum_{i=1}^{n} b_{ki}a_{ik} = \text{tr}\left(\sum_{i=1}^{n} b_{ki}a_{ij}\right) \\ &= \text{tr}(BA). \end{aligned}$$

Consequently, we have the following important results:

(13.3) PROPOSITION. If A is nonsingular and A and B are n-by-n matrices, then $\text{tr}(A^{-1}BA) = \text{tr}(B)$.

PROOF. By (13.2),

$$\text{tr}(A^{-1}(BA)) = \text{tr}((BA)A^{-1}) = \text{tr}(B(AA^{-1})) = \text{tr}(B).$$

(13.4) PROPOSITION. If T and T^* are equivalent representations of G, then their associated characters χ and χ^* are equal as functions.

PROOF. Since T is equivalent to T^*, there exists a nonsingular matrix A such that $AT(g) = T^*(g)A$ for every $g \in G$. Then if $\chi(g) = \text{tr}(T(g))$ and $\chi^*(g) = \text{tr}(T^*(g))$ for each $g \in G$, we have

$$\chi(g) = \text{tr}(T(g)) = \text{tr}(A^{-1}T^*(g)A) = \text{tr}(T^*(g)) = \chi^*(g)$$

by (13.3), for every $g \in G$.

Note that proposition (13.4) specifies that χ and χ^* are equal *as functions*, that is, they have the same domain (namely, G) and the same value at each given element of that domain. Now we observe an interesting property of characters:

(13.5) PROPOSITION. If χ is a character of G, then χ is a *class function* on G, that is, if x and y are conjugate in G, then $\chi(x) = \chi(y)$.

PROOF. By hypothesis, there exists $g \in G$ such that $g^{-1}xg = y$. Let T be the representation associated with χ. Then

$$\begin{aligned} \chi(y) &= \text{tr}(T(y)) = \text{tr}(T(g^{-1}xg)) = \text{tr}(T(g^{-1})T(x)T(g)) \\ &= \text{tr}(T(g)^{-1}T(x)T(g)) = \text{tr}(T(x)) = \chi(x), \end{aligned}$$

using (13.3) at the next to last equivalence.

If T is reducible, then T is equivalent to a representation T^* of the form

$$T^*(g) = \begin{bmatrix} U(g) & Q(g) \\ \bar{0} & V(g) \end{bmatrix}, \qquad g \in G,$$

where U and V are representations of G and each $Q(g)$ is a submatrix of the appropriate size. Then, clearly, the character of T^* is equal to the sum of the characters of U and V, and by (13.4), the character of T is also equal to the sum of the characters of U and V.

Conversely, just as we can form a new representation

$$T(g) = \begin{bmatrix} U(g) & \bar{0} \\ \bar{0} & V(g) \end{bmatrix}, \qquad g \in G,$$

from given representations U and V of G, so we can form a new character χ given characters λ and μ of G by setting

$$\chi(g) = \lambda(g) + \mu(g) \quad \text{for all} \quad g \in G.$$

A word of caution is in order. A representation is a homomorphism, and we have consistently used the fact that if T is a representation of G, then $T(xy) = T(x)T(y)$ for any x, $y \in G$. But the trace function does not in general preserve products, and thus, if χ is the character associated with T, $\chi(xy)$ will not in general be equal to $\chi(x)\chi(y)$. Even when $y = x$, we will not ordinarily have $\chi(x^2)$ equal to $\chi(x)^2$, as the example

$$T(x) = \begin{bmatrix} 0 & 1 \\ 1 & 0 \end{bmatrix}$$

for $\langle x : x^2 = 1 \rangle$ shows. Here $\chi(x) = 0$, but $\chi(x^2) = \chi(1) = 2$.

Although I omit the proof, it is true that the converse of (13.4) holds when G is finite:

(13.6) THEOREM. Two representations of a *finite* group are equivalent if their characters are equal as functions.

We now return to recalling ideas from linear algebra. Let τ be a linear transformation from $\mathbf{C}^n$ to itself. If X is a vector (in $\mathbf{C}^n$) and λ a (complex) scalar such that

$$\tau(X) = \lambda X, \tag{13.7}$$

then X is called a *characteristic vector* of τ and λ a *characteristic root* of τ. Sometimes the terms *eigenvector* and *eigenvalue* are used. If A is the matrix for τ, that is, if $\tau(X)=AX$ for each $X\in\mathbf{C}^n$, then X and λ satisfying (13.7) are also called a characteristic vector and characteristic root *of the matrix* A. If A is an n-by-n matrix, then one shows in linear algebra that A has n characteristic roots (not necessarily distinct), which are the n roots of the *characteristic polynomial*

$$\det(I_n x - A)$$

of A, where x is regarded as the variable of the polynomial. When no confusion with our notation for the order of a group will result, we shall continue to denote the determinant of a square matrix M by $|M|$, as we did in section 11.

For example, let

$$A=\begin{bmatrix}1 & 2\\ 3 & 2\end{bmatrix}.$$

Then

$$|I_2 x - A| = \begin{vmatrix} x-1 & -2\\ -3 & x-2\end{vmatrix} = (x-1)(x-2)-6$$
$$= x^2-3x-4=(x-4)(x+1),$$

so A has characteristic roots -1 and 4. We can then solve (13.7) using -1 for λ:

$$AX=-X$$
$$\begin{bmatrix}1 & 2\\ 3 & 2\end{bmatrix}\begin{bmatrix}x\\ y\end{bmatrix}=\begin{bmatrix}-x\\ -y\end{bmatrix}$$
$$\begin{bmatrix}x+2y\\ 3x+2y\end{bmatrix}=\begin{bmatrix}-x\\ -y\end{bmatrix}$$

from which $x+2y=-x$ yields $2y=-2x$, or $x=-y$ (and $3x+2y=-y$ also gives $x=-y$). This tells us that any vector with $x=-y$ is an eigenvector associated with $\lambda=-1$; in particular, we could choose

$$X=\begin{bmatrix}1\\ -1\end{bmatrix}.$$

Indeed, if $AX=\lambda X$ and if c is any scalar, then

$$A(cX)=c(AX)=c(\lambda X)=\lambda(cX),$$

so cX is also an eigenvector of A. Now if $c=0$, the observation is correct but trivial since a linear transformation always carries the zero vector to itself. Hence we say for the given A that *any nonzero multiple* of

$$\begin{bmatrix} 1 \\ -1 \end{bmatrix}$$

is an eigenvector associated with $\lambda = -1$.

If we solve (13.7) using $\lambda = 4$, we set $AX = 4X$, and a computation similar to the one above shows that any nonzero multiple of

$$\begin{bmatrix} 2 \\ 3 \end{bmatrix}$$

is an eigenvector of the given A associated with $\lambda = 4$.

I have used column vectors here because they are probably more familiar from discussions of characteristic roots and vectors in linear algebra. You can easily check that if we use a row vector X and write $\tau(X) = XA = \lambda X$, then the eigenvalues are the same (which comes from the observation that $|I_n x - A^t|$ is equal to $|(I_n x - A)^t|$) but that for the given A, the eigenvectors are any nonzero multiples of $(3, -2)$ and $(1, 1)$ for $\lambda = 4, -1$, respectively. The fact that the transposes of these vectors are perpendicular to the column eigenvectors found for A is purely coincidental, as may be checked by finding the row and column eigenvectors for

$$B = \begin{bmatrix} 1 & 3 \\ 2 & 2 \end{bmatrix}$$

and $\lambda = -1$. Since we shall need to work only with eigenvalues and not with eigenvectors in this section, the choice between rows and columns is immaterial.

If $\alpha_1, \ldots, \alpha_n$ are the characteristic roots of a matrix A, then

$$(13.8) \quad \begin{cases} |I_n x - A| = (x - \alpha_1)(x - \alpha_2) \ldots (x - \alpha_n) \\ \qquad = x^n - \left(\sum_{j=1}^{n} \alpha_j\right) x^{n-1} + \{\text{terms in } x^{n-2}, \text{etc.}\}. \end{cases}$$

Now we expand $|I_n x - A|$ by minors of the first row:

$$(13.9) \quad \begin{cases} |I_n x - A| = \begin{vmatrix} x - a_{11} & -a_{12} & \ldots & -a_{1n} \\ -a_{21} & x - a_{22} & \ldots & -a_{2n} \\ \ldots & \ldots & & \ldots \\ -a_{n1} & -a_{n2} & \ldots & x - a_{nn} \end{vmatrix} \\ \qquad = (x - a_{11}) \begin{vmatrix} x - a_{22} & \ldots & -a_{2n} \\ \ldots & & \ldots \\ -a_{n2} & \ldots & x - a_{nn} \end{vmatrix} + Q \end{cases}$$

where Q consists of terms involving only $(n-2)$ entries of the form $(x-a_{jj})$. Hence the coefficients of x^n and x^{n-1} in the characteristic polynomial must come only from the first term of the second line of (13.9). If we continue the process and expand that first term by minors (or use mathematical induction), we arrive at

(13.10) $$|I_n x - A| = (x-a_{11})(x-a_{22})\dots(x-a_{nn}) + Q^*,$$

where Q^* consists of terms in x^{n-2}, x^{n-3}, . . . , x, and constants. Now expansion of (13.10) gives

$$(x-a_{11})\dots(x-a_{nn}) = x^n - \left(\sum_{j=1}^{n} a_{jj}\right)x^{n-1} + Q^{**},$$

where Q^{**} consists of terms in x^{n-2}, etc. Comparing the coefficients of x^{n-1} in this last formula with those in the second line of (13.8), we see that

$$\sum_{j=1}^{n} \alpha_j = \sum_{j=1}^{n} a_{jj},$$

that is,

(13.11) PROPOSITION. The trace of a square matrix A equals the sum of the characteristic roots of A.

Now as a consequence of the definitions involved, we can show:

(13.12) PROPOSITION. Let A be an n-by-n matrix with characteristic roots $\alpha_1, \dots, \alpha_n$, and let k be a positive integer. Then the characteristic roots of A^k are $\alpha_1^k, \dots, \alpha_n^k$.

PROOF. Let $1 \le j \le n$. Then there exists a vector X_j such that $AX_j = \alpha_j X_j$. But then

$$\begin{aligned} A^k X_j &= A^{k-1}(\alpha_j X_j) = \alpha_j A^{k-1} X_j \\ &= \alpha_j A^{k-2}(\alpha_j X_j) = \alpha_j^2 A^{k-2} X_j, \end{aligned}$$

and by repeating this process we arrive at

$$A^k X_j = \alpha_j^k X_j,$$

so α_j^k is a characteristic root of A^k (with associated eigenvector X_j). Since A^k has only n characteristic roots, these roots must be α_j^k for $1 \le j \le n$. (Of course the same proof could be written with row vectors on the left of A rather than column vectors on the right.)

A deeper but very important result is:

(13.13) THEOREM. If $AB = BA$, and if the characteristic roots of A and B are $\alpha_1, \dots, \alpha_n$ and $\beta_1, \dots, \beta_n$, respectively, then the characteristic roots of AB are

$$\alpha_1\beta_1, \ldots, \alpha_n\beta_n,$$

where the β_j have been renumbered as may be needed. (See Curtis and Reiner [**4**], p. 208.)

Although I omit the proof of (13.13), we shall prove some useful consequences.

(13.14) COROLLARY. If A is a nonsingular matrix with characteristic roots $\alpha_1, \ldots, \alpha_n$, then A^{-1} has characteristic roots $1/\alpha_1, \ldots, 1/\alpha_n$.

PROOF. Since $AA^{-1} = A^{-1}A[= I_n]$, the theorem applies. Now the characteristic roots of I_n are all equal to 1 since the characteristic polynomial is

$$|I_n x - I_n| = |(x-1)I_n| = (x-1)^n,$$

and $(x-1)^n$ has only one root, namely 1, counted n times. If $\beta_1, \ldots, \beta_n$ are the characteristic roots of A^{-1}, then (by changing the order of the indices as may be needed)

$$\alpha_1\beta_1 = \alpha_2\beta_2 = \ldots = \alpha_n\beta_n = 1.$$

Hence $\beta_j = 1/\alpha_j$ for each $j = 1, \ldots, n$.

(13.15) COROLLARY. If A is a matrix of finite order, then the characteristic roots of A are all roots of 1.

PROOF. Let k be the order of A, that is, k is the smallest positive integer such that $A^k = I_n$, where we recall that A is in the group $GL(n, \mathbf{C})$. Then by (13.12), the roots of A^k are the kth powers of the roots of A, but as in the proof of (13.14), these must all equal 1.

Recall that if $z = a + bi$ is a complex number with a, b real, then the *complex conjugate* (or simply *conjugate*) of z is $a - bi$, which is denoted by $\bar{z}$. Thus, we easily see:

(13.16) PROPOSITION. If $z_1, z_2 \in \mathbf{C}$, then

$$\overline{z_1 + z_2} = \overline{z_1} + \overline{z_2} \quad \text{and} \quad \overline{z_1 z_2} = \overline{z_1}\,\overline{z_2}.$$

Again for $z \in \mathbf{C}$, we define the *norm* of z to be $\sqrt{z\,\bar{z}}$, which we denote by $|z|$. Note that if z is real, then $|z| = \sqrt{z^2}$, which is the absolute value of z, and thus the notation $|z|$ is consistent. For any $z \in \mathbf{C}$, $|z|$ is the length of the vector representing z in the complex plane (that is, in an Argand diagram). Moreover:

(13.17) PROPOSITION. If $z_1, z_2 \in \mathbf{C}$, then

$$|z_1 z_2| = |z_1|\,|z_2|.$$

The proof is left to exercise 13.3. Furthermore, we have:

(13.18). PROPOSITION. If ζ is a kth root of 1, then $|\zeta| = 1$ and $\zeta^{-1} = \bar{\zeta}$.

PROOF. By $k-1$ applications of (13.17), we have

$$|\zeta|^k = |\zeta^k| = |1| = 1.$$

But clearly the norm of any complex number is a nonnegative real number, and the only nonnegative real root of 1 is 1 itself; hence $|\zeta| = 1$. Now since $\zeta\bar{\zeta} = |\zeta|^2 = 1^2 = 1$, we have $\bar{\zeta} = 1/\zeta = \zeta^{-1}$.

The next two results follow from theorem (13.13).

(13.19) COROLLARY. If T is a representation of G with character χ, and if $x \in G$ is an element of finite order, then $\chi(x^{-1}) = \overline{\chi(x)}$.

PROOF. Let k be the order of the matrix $T(x)$ in the group $GL(n, \mathbf{C})$. By (13.14) and (13.15), the characteristic roots of $T(x^{-1})$ are the inverses of those of $T(x)$, that is, the reciprocals, which are multiplicative inverses. By (13.18) the inverse of a root of 1 is its conjugate. But by (13.11), $\chi(x)$ and $\chi(x^{-1})$ are, respectively, the sums of the characteristic roots of $T(x)$ and of $T(x^{-1})$. By $k-1$ applications of (13.16),

$$\chi(x^{-1}) = \overline{\chi(x)}.$$

(13.20) COROLLARY. Let T be a representation of G with character χ, let $t = \text{degree}(T)$, and let $g \in G$ be an element of finite order. Then $|\chi(g)| \leq t$, with equality holding if and only if T is equivalent to some representation T^* with $T^*(x) = \lambda I_t$ for some $\lambda \in \mathbf{C}$. (This is a special case of corollary (30.11) in Curtis and Reiner [**4**], p. 212.)

PROOF. Let $\alpha_1, \ldots, \alpha_t$ be the characteristic roots of $T(g)$ and k be the order of g. (Note that the order of $T(g)$ *divides* k, but may be strictly less than k.) By (13.15), the α_j are all k^{th} roots of 1. By (13.11) and the Triangle Inequality,

$$|\chi(g)| = \left|\sum_{j=1}^{t} \alpha_j\right| \leq \sum_{j=1}^{t} |\alpha_j| = t \cdot 1 = t, \tag{13.21}$$

establishing the inequality. We show next that equality holds in (13.21) if and only if $\alpha_1 = \alpha_2 = \ldots = \alpha_t$. Recall that for any nonzero complex numbers z_1 and z_2,

$$(*) \qquad |z_1 + z_2| = |z_1| + |z_2|$$

if and only if there is a positive real number c such that $z_2 = cz_1$. (This is easily seen by taking z_1 and z_2 as vectors in the complex plane and considering their vector sum.) If $|z_1| = |z_2|$, then the equality in (*) holds if and only if $z_1 = z_2$. This establishes our claim in the case $t = 2$; we proceed by induction. If $\alpha_1 = \ldots = \alpha_t$, then certainly equality holds in (13.21). For the converse, let

$$\left| \sum_{i=1}^{t} \alpha_i \right| = t.$$

Then

$$\left| \sum_{i=1}^{t-1} \alpha_i \right| = t - 1$$

and $|\alpha_{t-1} + \alpha_t| = 2$ since otherwise inequality would hold in (13.21). Hence $\alpha_1 = \ldots = \alpha_{t-1}$ and $\alpha_{t-1} = \alpha_t$. Hence $\chi(g)$ is the trace of a matrix $T^*(g)$ that satisfies both of the polynomial equations

$$x^k - I_t = \bar{0} \quad \text{and} \quad (x - \alpha_1 I_t)^t = \bar{0}.$$

From the theory of rings of polynomials as developed in abstract algebra, we may consider these polynomials simply as

$$(13.22) \qquad x^k - 1 = 0 \quad \text{and} \quad (x - \alpha_1)^t = 0.$$

Now using complex coefficients,

$$x^k - 1 = (x - 1)(x - \zeta)(x - \zeta^2) \ldots (x - \zeta^{k-1}),$$

where ζ is a primitive k^{th} root of 1. Again, in studying rings of polynomials, one proves that $T^*(g)$ must have a unique monic polynomial $p(x)$ such that $p(T^*(g)) = \bar{0}$ and such that $p(x)$ divides every polynomial $f(x)$ for which $f(T^*(g)) = \bar{0}$. Now there exist complex numbers $\eta_1, \ldots, \eta_s$ such that

$$p(x) = (x - \eta_1)(x - \eta_2) \ldots (x - \eta_s),$$

where s is the degree of $p(x)$. Note that since $p(x)$ divides both polynomials in (13.22), we have $s \leq k$ and $s \leq t$. But $1, \zeta, \zeta^2, \ldots, \zeta^{k-1}$ are all distinct, so $x^k - 1$ and $(x - \alpha_1)^t$ can have only one factor in common, namely $(x-\alpha_1)$. Hence

$$p(x)=x-\alpha_1, \quad \text{that is,} \quad p(x)=x-\alpha_1 I_t,$$

and so $T^*(g)-\alpha_1 I_t=\bar{0}$, which gives $T^*(g)=\alpha_1 I_t$. Here α_1 is the complex scalar λ required by the conclusion. The equivalence of T and T^* (needed only on the subgroup $\langle g\rangle$) follows by (13.6).

In (13.19) and (13.20) we required merely that $g\in G$ be an element of finite order and not that G itself be finite. Of course, if G is finite, then all of its elements have finite order, and the hypothesis is satisfied for g.

We shall use crucially but without proof the following:

(13.23) THEOREM. The number of inequivalent irreducible representations of a finite group G is equal to the number of distinct conjugate classes of G. (For a proof, see Curtis and Reiner [**4**], p. 186.)

As our first application of this theorem, we prove:

(13.24) THEOREM. If G is a finite abelian group, then every irreducible representation of G has degree 1.

PROOF. Every element of G is *self-conjugate*, that is, if $x\in G$ then $g^{-1}xg=x$ for every $g\in G$. Hence each element of G forms its own conjugate class, and by (13.23) the number of inequivalent irreducible representations of G is $|G|$. But by (12.17), there are $|G|$ one-dimensional representations of G, and they certainly are irreducible. This completes the proof.

We return to the familiar example of D_4. Writing

$$D_4=\langle x, y : x^4=y^2=1,\ yx=x^{-1}y\rangle,$$

we can specify any k-dimensional representation of D_4 by giving $T(x)$ and $T(y)$ as k-by-k matrices and verifying that

$$T(x)^4=T(y)^2=I_k \quad \text{and} \quad T(y)T(x)=T(x)^{-1}T(y).$$

The one-dimensional representations of D_4 are to be found in exercise 13.4.

Now consider the representation of degree 2 given by

$$U(x)=\begin{bmatrix}0 & -1\\ 1 & 0\end{bmatrix}, \qquad U(y)=\begin{bmatrix}0 & 1\\ 1 & 0\end{bmatrix}.$$

Note that, as required, $U(x)^4 = U(y)^2 = I_2$ and

$$U(y)U(x) = \begin{bmatrix} 1 & 0 \\ 0 & -1 \end{bmatrix} = \begin{bmatrix} 0 & 1 \\ -1 & 0 \end{bmatrix}\begin{bmatrix} 0 & 1 \\ 1 & 0 \end{bmatrix} = U(x)^{-1}U(y).$$

To show that U is irreducible over **C**, we need (by Maschke's Theorem (11.11)) merely to show that it is indecomposable. Suppose that U is *decomposable*; then there exists

$$A = \begin{bmatrix} a & b \\ c & d \end{bmatrix}$$

with $ad - bc \neq 0$ and there are complex numbers λ, μ, α, β with

$$\text{(13.25)} \qquad \begin{cases} A^{-1}U(x)A = \begin{bmatrix} \lambda & 0 \\ 0 & \mu \end{bmatrix}, \\ A^{-1}U(y)A = \begin{bmatrix} \alpha & 0 \\ 0 & \beta \end{bmatrix}. \end{cases}$$

Taking determinants on both sides, we have

$$1 = \lambda\mu \quad \text{and} \quad -1 = \alpha\beta.$$

Taking traces on both sides, we have, by (13.3),

$$0 = \lambda + \mu \quad \text{and} \quad 0 = \alpha + \beta.$$

Hence $\lambda = \pm i$, $\mu = \mp i$, $\alpha = \pm 1$, $\beta = \mp 1$. Now from (13.25).

$$\begin{bmatrix} -c & -d \\ a & b \end{bmatrix} = \begin{bmatrix} \lambda a & \mu b \\ \lambda c & \mu d \end{bmatrix} \quad \text{and} \quad \begin{bmatrix} c & d \\ a & b \end{bmatrix} = \begin{bmatrix} \alpha a & \beta b \\ \alpha c & \beta d \end{bmatrix}.$$

Thus $\lambda a = -c$ and $\alpha a = c$, whence $-\lambda a = \alpha a$, that is, $\mp ia = \pm a$, which can only hold when $a = 0$. But $a = 0$ gives $c = 0$ and thus $\det(A) = 0$, a contradiction. Hence U is irreducible.

In summary, we tabulate the characters of all irreducible representations of D_4. We observe first (see exercise 13.5) that D_4 has precisely five conjugate classes; if we write the group as $\langle x, y : x^4 = y^2 = 1, yx = x^3y \rangle$, the conjugate classes are $\{1\}$, $\{x^2\}$, $\{x, x^3\}$, $\{y, x^2y\}$, $\{xy, x^3y\}$. Hence the irreducible representation U just found, together with the four one-dimensional representations in exercise 13.4, constitute all of the irreducible representations of D_4. If χ is the character associated with U, then $\chi(1) = 2$, $\chi(x) = 0$, $\chi(y) = 0$, $\chi(xy) = 0$, and $\chi(x^2) = -2$, all from the matrices above. Thus we have the following *character table* for D_4:

	Conjugate classes				
Characters	$\{1\}$	$\{x^2\}$	$\{x, x^3\}$	$\{y, x^2y\}$	$\{xy, x^3y\}$
χ_1	1	1	1	1	1
χ_2	1	1	−1	1	−1
χ_3	1	1	1	−1	−1
χ_4	1	1	−1	−1	1
χ_5	2	−2	0	0	0

EXERCISES

13.1. Exhibit 2-by-2 matrices A and B that show that $\operatorname{tr}(AB)$ need not equal $\operatorname{tr}(A)\operatorname{tr}(B)$, but verify that (13.2) still holds.

13.2. Find the characteristic values and vectors for

$$A = \begin{bmatrix} 1 & 2 \\ 8 & 1 \end{bmatrix}.$$

13.3. Prove (13.16); show that for $z \in \mathbf{C}$, the product $z\bar{z}$ is always real; and prove (13.17).

13.4. Verify that D_4, as given in the text, has precisely four representations of degree 1, given by $T(x) = [\pm 1]$ and $T(y) = [\pm 1]$. (*Hint:* This may be done inelegantly as we did it for S_3 in section 9, or more elegantly by using (12.18) through (12.22).)

13.5. Find the distinct conjugate classes of D_4. (*Hint:* For a given element, consider what happens to it under conjugation by the two generators of D_4, then what happens to the elements thus obtained, etc. Recall that conjugate classes are orbits, hence are disjoint.)

• *13.6.* Use (12.18) through (12.22) to find all of the one-dimensional representations of Q_2.

13.7. If G is a group that coincides with its derived group (that is, $G = G'$), what can be said about the one-dimensional representations (and hence characters) of G over $\mathbf{C}$?

• *13.8.* The alternating group A_4 of order 12 may be written as

$$\langle x, y : x^3 = y^3 = 1,\ yx = x^2y^2 \rangle.$$

It has only one nontrival normal subgroup, which is $\langle xy, x^2y^2 \rangle$. What is the derived group of A_4? (*Hint:* Use (12.20).) Find all of the one-dimensional representations (and characters) of A_4.

13.9. The group $G = \langle a, b : a^5 = b^4 = 1,\ ba = a^2b \rangle$ has order 20; its only nontrivial normal subgroup is $\langle a \rangle$. Find all of the one-dimensional representations of G.

Section 14

Orthogonality Relations and Character Tables

As we have seen, a character is a function from a group G into the complex numbers $\mathbf{C}$. If G is a *finite* group and if χ and ψ are *any* two functions from G into $\mathbf{C}$, we may define the *inner product* of χ and ψ by

$$\langle \chi, \psi \rangle = \frac{1}{|G|} \sum_{g \in G} \chi(g)\overline{\psi(g)}, \tag{14.1}$$

where the bar still represents complex conjugate, and the summation indicates that there is to be one addend for each element of G (which makes sense since G is finite). The fact that $\langle \chi, \psi \rangle$ is a complex number will be used crucially. The context of our discussions will always make clear whether the notation of pointed brackets is being used for the inner product of two functions or for a group (or subgroup) with two generators.

As an example, let $G = D_4$ and consider the characters computed in the table at the end of the last section. We have

$$\langle \chi_2, \chi_5 \rangle = \tfrac{1}{8}(2-2+0+0+0+0+0+0) = 0,$$
$$\langle \chi_2, \chi_2 \rangle = \tfrac{1}{8}(1+1+1+1+1+1+1+1) = 1.$$

Before continuing with our discussion, we should justify our reference to $\langle\, , \rangle$ as an inner product. To do so, we must verify that $\langle\, , \rangle$ is a complex-valued symmetric bilinear form. Let χ, ψ, and ϕ be functions from a finite group G to $\mathbf{C}$, and let $a, b \in \mathbf{C}$. The property of *bilinearity* is that

$$\left\{ \begin{array}{ll} & \langle a\chi + b\psi, \ \phi \rangle = a\langle \chi, \phi \rangle + b\langle \psi, \phi \rangle \\ \text{and} & \\ & \langle \ \chi, a\psi + b\phi \rangle = \bar{a}\langle \chi, \psi \rangle + \bar{b}\langle \chi, \phi \rangle. \end{array} \right. \tag{14.2}$$

This check is routine and has been left as exercise 14.8. The property of *symmetry* is that

$$\langle \chi, \psi \rangle = \overline{\langle \psi, \chi \rangle}. \tag{14.3}$$

(Note that the complex conjugate is taken when the order of the entries is reversed.) To check (14.3) we show first that if $z_1, z_2 \in \mathbf{C}$, then $z_1 \overline{z_2} = \overline{z_2 \overline{z_1}}$. Let $z_1 = a + bi$ and $z_2 = c + di$; then

$$\begin{aligned}(a+bi)\overline{(c+di)} &= (a+bi)(c-di)\\ &= (ac+bd)+(bc-ad)i\\ &= (ca+db)-(da-cb)i\\ &= \overline{(c+di)(a-bi)}\\ &= \overline{(c+di)\overline{(a+bi)}},\end{aligned}$$

as was to be shown. Hence

$$\begin{aligned}\langle \chi, \psi \rangle &= \frac{1}{|G|} \sum_{g \in G} \chi(g)\overline{\psi(g)}\\ &= \overline{\frac{1}{|G|} \sum_{g \in G} \psi(g)\overline{\chi(g)}}\\ &= \overline{\langle \psi, \chi \rangle}.\end{aligned}$$

When the functions χ and ψ in (14.1) are characters, we may use the fact (from (13.5)) that characters are class functions to simplify the expression in (14.1). Let G have the distinct conjugate classes $C_1, \ldots, C_s$, let C_j contain h_j elements, and let g_j be any element of C_j, for $1 \leq j \leq s$. Then if χ and ψ are characters of G, (14.1) becomes

$$\langle \chi, \psi \rangle = \frac{1}{|G|} \sum_{j=1}^{s} h_j \chi(g_j)\overline{\psi(g_j)}. \tag{14.4}$$

In the example of D_4, (14.4) requires only five terms in each sum, in comparison to the eight required by (14.1); greater simplifications will frequently come about in larger groups. Of course an abelian group has as many conjugate classes as elements, and in this case the two expressions are the same.

Let δ_{ij} denote the *Kronecker delta* defined by

$$\delta_{ij} = \begin{cases} 1 & \text{if } i = j, \\ 0 & \text{if } i \neq j. \end{cases}$$

If we carry out the 25 computations of $\langle \chi_i, \chi_j \rangle$ with $1 \leq i, j \leq 5$ for D_4, we find that $\langle \chi_i, \chi_j \rangle = \delta_{ij}$. In fact, this statement is always true, though it is stated here without proof. First we remark that if T is a representation of G that is irreducible over $\mathbf{C}$, and if χ is the character associated with T, we shall call χ an *irreducible character* of G.

(14.5) THEOREM. Let G be a finite group having the distinct irreducible characters $\zeta^{(1)}, \ldots, \zeta^{(s)}$. Then

$$\langle \zeta^{(i)}, \zeta^{(j)} \rangle = \delta_{ij}$$

for $1 \leq i, j \leq s$.

In this theorem distinct characters are denoted by means of superscripts (placed in parentheses to avoid confusion with powers). The conventional notation for the value of the character $\zeta^{(i)}$ on an element of the conjugate class C_j of G is $\zeta_j^{(i)}$, that is

(14.6) $$\zeta_j^{(i)} = \zeta^{(i)}(g_j), \quad \text{where} \quad g_j \in C_j.$$

Because of (13.23), we note that both i and j in (14.6) run over the integers $1, \ldots, s$, where s is the number of distinct conjugate classes of G.

In the notation of (14.6), we define a *character table* for a finite group G to be the matrix $(\zeta_j^{(i)})$, that is, the matrix whose rows represent the distinct irreducible characters of G and whose columns represent the distinct conjugate classes. By (13.23), this matrix is square. Note that the table at the end of section 13 is arranged as described here.

Given a character table for G, we define the *conjugate* of any column (or row) of the table to be the column (or row) vector whose entries are, respectively, the complex conjugates of the entries in the given column (or row). Then from (14.5) we have:

(14.7) COROLLARY. In a character table, let X be the row vector $(h_j\zeta_j^{(i)})$ and $\bar{Y}$ the conjugate of the row vector $(\zeta_j^{(k)})$. If $i \neq k$, then the ordinary dot product $X \cdot \bar{Y} = 0$.

PROOF. $X \cdot \bar{Y} = \sum_{j=1}^{s} (h_j\zeta_j^{(i)})\overline{\zeta_j^{(k)}} = |G|\langle \zeta^{(i)}, \zeta^{(k)} \rangle = |G|\delta_{ik} = 0.$

(14.8) COROLLARY. In a character table, let X be the row vector $(h_j\zeta_j^{(i)})$ and consider the row vector $(\zeta_j^{(i)})$. Then the ordinary dot product $X \cdot \overline{(\zeta_j^{(i)})} = |G|$.

Before applying these results to the construction of character tables, we must examine:

(14.9) THEOREM. Let G be a finite group having the distinct irreducible characters $\zeta^{(1)}, \ldots, \zeta^{(s)}$, and let $1 \leq i, j \leq s$.

Then

$$\sum_{k=1}^{s} h_i \zeta_i^{(k)} \overline{\zeta_j^{(k)}} = |G|\delta_{ij},$$

where h_i is the number of elements in the conjugate class C_i.

From this theorem, stated here without proof, we can derive three corollaries.

Note that the two conjugate classes (the ith and j^{th}) in (14.9) are fixed and that the sum is taken over the values of the irreducible characters on those classes. Moreover, since i is fixed, (14.9) is equivalent to

(14.10) $$\sum_{k=1}^{s} \zeta_i^{(k)} \overline{\zeta_j^{(k)}} = \frac{|G|}{h_i}\,\delta_{ij},$$

where h_i is merely a constant that appears in each term of the original sum. Now $\zeta_i^{(k)}$ is the value of $\zeta^{(k)}$ on elements of the ith conjugate class, which values are found in the ith column of the character table, and similarly for $\zeta_j^{(k)}$. Thus we have proved both of the following:

(14.11) COROLLARY. In a character table, the dot product of any column with the conjugate of any other column is 0.

(14.12) COROLLARY. In a character table, the dot product of the ith column with its own conjugate is $|G|/h_i$.

Furthermore, if we agree to take $C_1 = \{1\}$ for any group G, then the first column of the character table gives the degrees of the distinct irreducible representations (that is, the traces of the identity matrices whose degrees are the degrees of the representations). Now the degrees are all positive integers and so are equal to their own complex conjugates. Moreover, $h_1 = 1$. Hence from (14.12) we have proved:

(14.13) COROLLARY. The sum of the squares of the degrees of the distinct irreducible characters of G is equal to $|G|$.

We now consider some examples of groups to illustrate application of the five corollaries. For the first example, let G be S_3, the nonabelian group of order 6. We have precisely three conjugate classes, consisting of the elements of order 1, 2, and 3, respectively. By (14.13) we need integers z_1, z_2, z_3 such that $z_1^2 + z_2^2 + z_3^2 = 6$. A moment's reflection shows that two of these must equal 1 and the other 2. It is usual to list the irreducible characters in a table in order of increasing degree; hence we take $z_1 = z_2 = 1$ and $z_3 = 2$. Let C_i consist of the elements of order i, for $i = 1, 2, 3$. We certainly have the trivial representation

$$T_1(g) = [1] \quad \text{for all} \quad g \in G,$$

and it gives the *trivial character*

$$\zeta^{(1)}(g) = 1 \quad \text{for all} \quad g \in G.$$

From section 9 we know that $\zeta_2^{(2)} = -1$, $\zeta_3^{(2)} = 1$ is also the character of a one-dimensional representation of S_3. The information we have accumulated may be put into the character table as

	C_1	C_2	C_3
$\zeta^{(1)}$	1	1	1
$\zeta^{(2)}$	1	-1	1
$\zeta^{(3)}$	2		

Now if we compare the second column with the first and then the third with the first, by (14.11) we obtain $\zeta_2^{(3)} = 0$ and $\zeta_3^{(3)} = -1$.

Next we consider the group Q_2 of order 8. One finds by computation that the conjugate classes are

$$C_1 = \{1\}, \quad C_2 = \{-1\}, \quad C_3 = \{\pm i\}, \quad C_4 = \{\pm j\}, \quad C_5 = \{\pm k\}.$$

The only solution to

$$\sum_{m=1}^{5} z_m^2 = 8$$

for integers z_m with $z_1 \leq z_2 \leq \ldots \leq z_5$ is

$$z_1 = z_2 = z_3 = z_4 = 1, \qquad z_5 = 2.$$

From exercise 13.6 we have the one-dimensional characters of Q_2, and we may summarize the information we have thus far as:

	C_1	C_2	C_3	C_4	C_5
$\zeta^{(1)}$	1	1	1	1	1
$\zeta^{(2)}$	1	1	-1	1	-1
$\zeta^{(3)}$	1	1	1	-1	-1
$\zeta^{(4)}$	1	1	-1	-1	1
$\zeta^{(5)}$	2				

Then by (14.11), the last row must read $(2\ -2\ 0\ 0\ 0)$ in order for the second through fifth columns (considered as vectors) to be orthogonal to the first column.

Note that Q_2 has the same character table as D_4, although the two groups are not isomorphic.

Now we consider A_4 from section 7 (or exercise 13.8); it happens that A_4 is isomorphic to the subgroup K of exercise 7.12. The group A_4 consists of all the even permutations on four points and has order 12. The conjugate classes are:

$$C_1 = \{e\},$$
$$C_2 = \{(12)(34), (13)(24), (14)(23)\},$$
$$C_3 = \{(123), (142), (134), (243)\},$$
$$C_4 = \{(132), (124), (143), (234)\}.$$

Now $H = C_1 \cup C_2$ is a normal subgroup of A_4 (since conjugation preserves the orders of elements, and H is a subgroup consisting of all the elements of orders 1 and 2), and $A_4/H \cong Z_3$, so the one-dimensional representations of Z_3 give representations of A_4 (exercise 13.8). Thus $z_1 = z_2 = z_3 = 1$, and hence $z_4 = 3$. (Alternatively, one can check by trial and error that 12 can be written as the sum of four positive squares in only the one way $1^2 + 1^2 + 1^2 + 3^2$.) By (14.11) we obtain the following character table:

	C_1	C_2	C_3	C_4
$\zeta^{(1)}$	1	1	1	1
$\zeta^{(2)}$	1	1	ζ	ζ^2
$\zeta^{(3)}$	1	1	ζ^2	ζ
$\zeta^{(4)}$	3	-1	0	0

where ζ is a primitive cube root of 1. In checking orthogonality of the columns we use the identity

$$1 + \zeta + \zeta^2 = \frac{1 - \zeta^3}{1 - \zeta} = \frac{0}{1 - \zeta} = 0.$$

For a more difficult example, we look at S_4, which has order 24. The derived group of S_4 is A_4 and $[S_4 : A_4] = 2$, so by (12.22), S_4 has exactly two one-dimensional characters, one of which is of course the trivial character, which we list as $\zeta^{(1)}$. By computation one finds that S_4 has five conjugate classes (see exercise 14.4), which may be tabulated as

i	h_i	Description of C_i
1	1	$\{e\}$
2	6	elements of the form $(\alpha\beta)$
3	8	elements of the form $(\alpha\beta\gamma)$
4	6	elements of the form $(\alpha\beta\gamma\delta)$
5	3	elements of the form $(\alpha\beta)(\gamma\delta)$

where α, β, γ, δ run over the points 1, 2, 3, 4 and are distinct. From (14.13) we find that the remaining three irreducible characters have degrees 2, 3, 3, respectively.

The subgroup $H = C_1 \cup C_5$ is normal in S_4, as it was in A_4 (see exercise 14.6). Now H does not contain the derived group A_4 of S_4; hence by (12.20) S_4/H is nonabelian. Since $[S_4:H]=6$, by (7.2), $S_4/H \cong S_3$. Now by (12.18) and the character table for S_3 we have the character $\zeta^{(3)}$ of degree 2. Incorporating what we have found thus far into the table, we have.

	C_1	C_2	C_3	C_4	C_5
$\zeta^{(1)}$	1	1	1	1	1
$\zeta^{(2)}$	1	-1	1	-1	1
$\zeta^{(3)}$	2	0	-1	0	2
$\zeta^{(4)}$	3				
$\zeta^{(5)}$	3				

Now applying (14.11) to columns 2 and 1, we see that $\zeta_2^{(4)} = -\zeta_2^{(5)}$. Similarly, $\zeta_4^{(4)} = -\zeta_4^{(5)}$ and $\zeta_3^{(4)} = -\zeta_3^{(5)}$. For columns 5 and 1 we have

$$1\cdot 1 + 1\cdot 1 + 2\cdot 2 + 3\zeta_5^{(4)} + 3\zeta_5^{(5)} = 0,$$

whence $\zeta_5^{(5)} = -2 - \zeta_5^{(4)}$. We may summarize these results using unknowns t, u, v, w in the fourth and fifth rows of the table as follows:

$\zeta^{(4)}$	3	t	u	v	w
$\zeta^{(5)}$	3	$-t$	$-u$	$-v$	$-2-w$.

Now (14.7) gives

$$\sum_{j=1}^{5} h_j \zeta_j^{(4)} \overline{\zeta_j^{(1)}} = 0 \quad \text{and} \quad \sum_{j=1}^{5} h_j \zeta_j^{(4)} \overline{\zeta_j^{(2)}} = 0$$

and these two equations yield upon substitution,

$$(14.14) \qquad \begin{cases} 3+6t+8u+6v+3w=0 \\ 3-6t+8u-6v+3w=0 \end{cases}$$

and the sum of these two is $6+16u+6w=0$, or $3+8u+3w=0$. But (14.7) also gives

$$\sum_{j=1}^{5} h_j \zeta_j^{(4)} \overline{\zeta_j^{(3)}} = 0,$$

whence $6+0-8u+0+6w=0$, or $3-4u+3w=0$. Solving our two equations (subtract the second from the first) yields $u=0$ and $w=-1$.

Now we apply (14.11) to columns 2 and 4:

$$1 \cdot 1 + (-1)(-1) + 0 \cdot 0 + t \cdot \bar{v} + (-t)(\overline{-v}) = 0,$$

whence $t\bar{v}=-1$. Next substitute $u=0$ and $w=-1$ into the first equation of (14.14); the result is $t+v=0$. Moreover, since the elements of C_2 have order 2, from the proof of (13.15) we know that t must be a sum of square roots of 1; thus t is an integer. Now the pair

$$t\bar{v}=-1, \qquad t+v=0$$

has *two* solutions, namely, $t=\pm 1$ and $v=\mp 1$. But note that since $w=-1$, we have $\zeta_5^{(5)}=-2-w=-1$ also, and thus the fourth and fifth rows of the table will be identical except in columns 2 and 4. Thus the two solutions for t and v merely correspond to permuting $\zeta^{(4)}$ and $\zeta^{(5)}$, and we may then set $t=1$ and $v=-1$ arbitrarily. This completes the following character table for S_4.

	C_1	C_2	C_3	C_4	C_5
$\zeta^{(1)}$	1	1	1	1	1
$\zeta^{(2)}$	1	−1	1	−1	1
$\zeta^{(3)}$	2	0	−1	0	2
$\zeta^{(4)}$	3	1	0	−1	−1
$\zeta^{(5)}$	3	−1	0	1	−1

Our final example is one that we shall not be able to complete until section 17, but it will be useful to start it here and to add to it in section 15. Consider the group A_5 of order 60, which was discussed at the end of section 7. Since A_5 is simple and nonabelian, (12.20) shows that the derived group A_5' is equal to A_5; hence by (12.22), A_5 has only one irreducible character of degree 1, namely, the trivial character.

By computation one can find that the conjugate classes of A_5 may be tabulated as follows:

i	h_i	Description of C_i
1	1	$\{e\}$
2	15	elements of the form $(\alpha\beta)(\gamma\delta)$
3	20	elements of the form $(\alpha\beta\gamma)$
4	12	conjugates of (12345)
5	12	conjugates of (12354)

Hence we know to look for five irreducible characters, only one of which has degree 1, but thus far we have no information about the other irreducible characters. In the next section we shall find an irreducible character of degree 4; then by (14.13), the sum of the degrees of the three remaining irreducible characters must be $60-(1^2+4^2)=43$. Obviously the maximum degree is 6 (since $7^2=49$ is already too large), and in fact one can check easily that $a^2+b^2+c^2=43$ has as integer solutions only 3, 3, and 5.

EXERCISES

14.1. From the character table at the end of section 13, compute $\langle\chi_2, \chi_3\rangle$ and $\langle\chi_5, \chi_5\rangle$.

14.2. Verify that (14.7), (14.8), (14.11), (14.12), and (14.13) hold for the character table of D_4 found at the end of section 13.

14.3. Note that two distinct conjugate classes of A_4 unite into a single conjugate class of S_4. Why is this possible?

14.4. Once one knows what the conjugate classes of S_4 are, it is not difficult to write down a derivation of them. Consider, for example, C_2. Given two distinct 2-cycles, they will either have one point in common or no point in common, that is, we need consider only the two pairs $(\alpha\beta)$, $(\alpha\gamma)$ and $(\alpha\beta)$, $(\gamma\delta)$. Show that any two 2-cycles are conjugate, that is, find elements x and y in terms of α, β, γ, δ such that $x^{-1}(\alpha\beta)x=(\alpha\gamma)$ and $y^{-1}(\alpha\beta)y=(\gamma\delta)$. Then show that no element of order 2 that is not a 2-cycle can be a conjugate of $(\alpha\beta)$. Finally, to find h_2, consider how many distinct elements of S_4 have the form $(\alpha\beta)$.

14.5. A subgroup H of a group G is called a *characteristic* subgroup if, whenever σ is an automorphism of G, $\sigma(H)=H$. Prove that a characteristic subgroup is always normal, but show by example that a normal subgroup need not be characteristic.

14.6. Let $H \leq G$ and assume that H contains all of the elements of G having one or more specified orders, and no other elements. Prove that H is a characteristic subgroup of G. (*Caution:* The hypothesis that $H \leq G$ is crucial!)

14.7. Prove that the center $C(G)$ and the derived group G' are characteristic subgroups of G. (See exercise 5.15.)

14.8. Verify equations (14.2). (*Note:* The second may be deduced from the first by (14.3).)

Section 15
Reducible Characters

By Maschke's Theorem (11.11), every complex representation of a finite group G is completely reducible. If T is any complex representation of G and χ the character of T, then T is equivalent to a representation of the form

$$\begin{bmatrix} T_1(g) & & & & \bar{0} \\ & T_2(g) & & & \\ & & \cdot & & \\ & & & \cdot & \\ & & & & \cdot \\ \bar{0} & & & & T_r(g) \end{bmatrix}$$

where $T_1, \ldots, T_r$ are irreducible and $g \in G$. Hence by (13.3), χ is equal to a linear combination of the irreducible characters $\zeta^{(1)}, \ldots, \zeta^{(s)}$ of G with nonnegative integers as coefficients, that is, given χ, there exist integers $a_1, \ldots, a_s \geq 0$ such that

$$\chi = \sum_{i=1}^{s} a_i \zeta^{(i)} \tag{15.1}$$

in the sense of the sum of functions, namely,

$$\chi(g) = \sum_{i=1}^{s} a_i \zeta^{(i)}(g) \quad \text{for} \quad g \in G.$$

For example, we found that the character of the regular representation (11.5) of V_4 is equal to the sum of the four irreducible characters of V_4 from (11.16).

Let χ be a character of G satisfying (15.1). Then for fixed j,

$$\langle \chi, \zeta^{(j)} \rangle = \frac{1}{|G|} \sum_{k=1}^{s} h_k \chi(g_k) \overline{\zeta_k^{(j)}}$$

in the notation of (14.4). Now this expression becomes

$$\begin{aligned}\langle \chi, \zeta^{(j)}\rangle &= \frac{1}{|G|}\sum_{k=1}^{s} h_k\left(\sum_{i=1}^{s} a_i\zeta_k^{(i)}\right)\overline{\zeta_k^{(j)}}\\ &= \left\langle \sum_{i=1}^{s} a_i\zeta^{(i)}, \zeta^{(j)}\right\rangle\\ &= \sum_{i=1}^{s} a_i\langle \zeta^{(i)}, \zeta^{(j)}\rangle\\ &= \sum_{i=1}^{s} a_i\,\delta_{ij} \quad \text{by (14.5)}\\ &= a_j.\end{aligned}$$

Thus we have $\langle \chi, \zeta^{(j)}\rangle = a_j$ for each $j = 1, \ldots, s$, and (15.1) now becomes

(15.2) $$\chi = \sum_{i=1}^{s} \langle \chi, \zeta^{(i)}\rangle \zeta^{(i)}.$$

Note in particular that although $\langle \chi, \zeta\rangle$ is generally just a complex number, $\langle \chi, \zeta^{(i)}\rangle$ will be a nonnegative integer.

The result (15.2) is so powerful that it can be used to reduce an arbitrary character χ to the sum of the irreducible characters of which χ is composed. We call $a_i = \langle \chi, \zeta^{(i)}\rangle$ the *multiplicity* of $\zeta^{(i)}$ in χ; if $a_i > 0$, $\zeta^{(i)}$ is called a *constituent* of χ.

As our first example, we let G be any finite group and χ the character of the regular representation of G. As we observed at the beginning of section 13, $\chi(1) = |G|$ and $\chi(g) = 0$ for all $g \neq 1$. Hence $\chi(g_i) = 0$ for every $g_i \in C_i$ with $i \geq 2$, where as usual $C_1 = \{1\}$, and $C_2, \ldots, C_s$ are the remaining distinct conjugate classes of G. Hence

$$\begin{aligned}\langle \chi, \zeta^{(i)}\rangle &= \frac{1}{|G|}\sum_{j=1}^{s} h_j\chi(g_j)\overline{\zeta_j^{(i)}}\\ &= \frac{1}{|G|}\chi(1)\overline{\zeta_1^{(i)}}\\ &= \frac{1}{|G|}\cdot |G| \cdot z_i\\ &= z_i,\end{aligned}$$

where z_i is the degree of $\zeta^{(i)}$. Thus we have proved:

(15.3) PROPOSITION. If G is a finite group and χ the character of the regular representation of G, then the multiplicity of each irreducible character of G in χ is equal to the degree of that character.

Note that (15.3) echoes (14.13): the degree of $\zeta^{(i)}$ and its multiplicity are equal, their product is z_i^2, and $\sum_{i=1}^{s} z_i^2$ equals $|G|$, the degree of the regular representation.

In exercise 9.7 we found a representation of D_4 given by

$$U(r)=\begin{bmatrix}0 & -1 & 0 & 0\\ 1 & 0 & 0 & 0\\ 0 & 0 & 0 & -1\\ 0 & 0 & 1 & 0\end{bmatrix},\qquad U(c)=\begin{bmatrix}0 & 1 & 1 & 0\\ 1 & 0 & 0 & -1\\ 0 & 0 & 0 & 1\\ 0 & 0 & 1 & 0\end{bmatrix}.$$

If χ is the character of U, then $\chi(r)=0$ and $\chi(c)=0$. Remember that we *cannot* find $\chi(r^2)$ and $\chi(rc)$ as the products $\chi(r)^2$ and $\chi(r)\chi(c)$. But by matrix multiplication we obtain

$$U(r^2)=\begin{bmatrix}-1 & 0 & 0 & 0\\ 0 & -1 & 0 & 0\\ 0 & 0 & -1 & 0\\ 0 & 0 & 0 & -1\end{bmatrix},\qquad U(rc)=\begin{bmatrix}-1 & 0 & 0 & 1\\ 0 & 1 & 1 & 0\\ 0 & 0 & -1 & 0\\ 0 & 0 & 0 & 1\end{bmatrix},$$

whence $\chi(r^2)=-4$ and $\chi(rc)=0$. Comparison with the table at the end of section 13 makes it clear that $\chi=2\zeta^{(5)}$, but it may be helpful to verify directly that $\langle\chi, \zeta^{(i)}\rangle=0$ for $i=1, 2, 3, 4$. (See exercise 15.1.)

For a less trivial (if unmotivated) example, suppose we have for S_4 the following character:

	C_1	C_2	C_3	C_4	C_5
χ	19	5	−2	−3	3

Then, using the character table for S_4 found at the end of section 14, we find

$$\langle\chi, \zeta^{(1)}\rangle=\tfrac{1}{24}(19+30-16-18+9)=1,$$
$$\langle\chi, \zeta^{(2)}\rangle=\tfrac{1}{24}(19-30-16+18+9)=0,$$
$$\langle\chi, \zeta^{(3)}\rangle=\tfrac{1}{24}(38+0+16+0+18)=3,$$
$$\langle\chi, \zeta^{(4)}\rangle=\tfrac{1}{24}(57+30+0+18-9)=4,$$
$$\langle\chi, \zeta^{(5)}\rangle=\tfrac{1}{24}(57-30+0-18-9)=0;$$

whence $\chi=\zeta^{(1)}+3\zeta^{(3)}+4\zeta^{(4)}$, which may easily be verified.

A more interesting example for S_4 may be derived from the group of rigid symmetries of the cube once we prove the following result:

(15.4) PROPOSITION. The group of rigid symmetries of the cube is isomorphic to S_4.

PROOF. Let the vertices of a cube be numbered as shown in figure 9, and denote the four principal diagonals (that is, the interior diagonals of length $\sqrt{3}$ times that of an edge) as

$$a = \overline{17}, \qquad b = \overline{28}, \qquad c = \overline{35}, \qquad d = \overline{46}.$$

Because we consider only rigid symmetries of the cube, any such motion permutes the four diagonals among themselves, and since $|G| = 24$, all of the permutations of the set $\{a, b, c, d\}$ must be included. This assertion depends upon the observation that no two elements of the symmetry group G induce the same permutation of the diagonals, which may be verified using (3.11) as follows. Since the orbit of vertex 1 under G contains all eight points, the orbit of the diagonal a must have four elements. The stabilizer of a in G will be the union of the stabilizer of the vertex 1 (which we know to have order 3) and the set of elements of G that carry vertex 1 to vertex 7 (reversing the endpoints of a). The latter subset consists of those elements that carry vertex 2 to a point adjacent to 7, that is, to 3, 6, or 8. Hence $|G_a| = 6$ and the order of the permutation group on the diagonals is 24; thus G is all of S_4. This much suffices to establish that the two groups are isomorphic, but we can make the isomorphism explicit by comparing the action

Figure 9

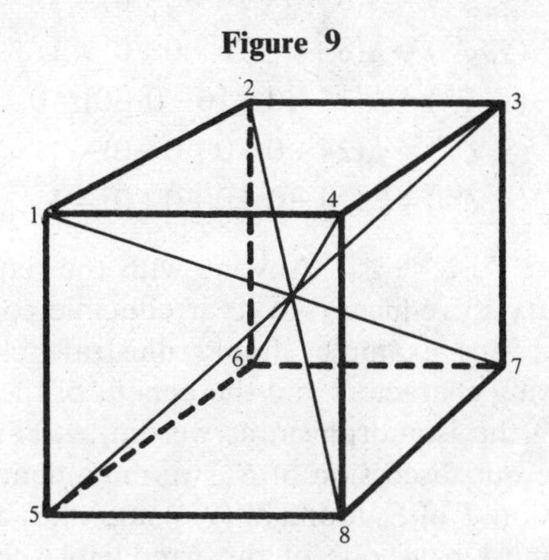

of a symmetry operation on the vertices and on the diagonals, for example:

$$(15.5)\begin{cases} r=(1234)(5678) & \text{corresponds to} \quad (abcd)\in C_4, \\ h=(245)(386) & \text{corresponds to} \quad (bdc)\in C_3, \\ r^2=(13)(24)(57)(68) & \text{corresponds to} \quad (ac)(bd)\in C_5, \\ rh=(14)(28)(35)(67) & \text{corresponds to} \quad (ad)\in C_2, \end{cases}$$

where we have denoted the conjugate classes of S_4 as at the end of section 14.

Now we can produce a representation of S_4 of degree 8 by letting the natural basis vectors $e_1, \ldots, e_8$ of $\mathbf{R}^8$ correspond to the vertices $1, \ldots, 8$, respectively, and by taking $T(g)$ so that $e_iT(g)= e_j$ if and only if $g: i \to j$, for each g in the symmetry group G. Let χ be the character associated with T, choose $g_i \in C_i$ for each conjugate class C_i of G, and let g be any one of these g_i $(1\leq i\leq 5)$. Since to find $\chi(g)$ we need to know only the entries on the main diagonal of the matrix $T(g)$, we may observe that $T(g)$ has a 1 on the main diagonal for each vertex left fixed by g, and a 0 on the main diagonal for each vertex moved by g. By observing the examples given in (15.5), we immediately produce the character

	C_1	C_2	C_3	C_4	C_5
χ	8	0	2	0	0

Now to reduce χ to a sum of irreducible characters, we compute

$$\begin{aligned} \langle\chi, \zeta^{(1)}\rangle &= \tfrac{1}{24}(8+0+16+0+0)=1, \\ \langle\chi, \zeta^{(2)}\rangle &= \tfrac{1}{24}(8+0+16+0+0)=1, \\ \langle\chi, \zeta^{(3)}\rangle &= \tfrac{1}{24}(16+0-16+0+0)=0, \\ \langle\chi, \zeta^{(4)}\rangle &= \tfrac{1}{24}(24+0+0+0+0)=1, \\ \langle\chi, \zeta^{(5)}\rangle &= \tfrac{1}{24}(24+0+0+0+0)=1; \end{aligned}$$

hence $\chi=\zeta^{(1)}+\zeta^{(2)}+\zeta^{(4)}+\zeta^{(5)}$. Anyone with the requisite time and practice could try to reduce T to its irreducible components as in section 11, but this example should illustrate clearly both the advantage of using characters and the benefit of the abstract group theory (utilizing the isomorphism as we did).

To conclude our discussion of S_4, we may identify the irreducible representation T of S_4 found at (9.4) and (9.5) and discussed in section 11. It must have one of the irreducible characters in the table at the end of section 14. Let ψ be the character of T. Then

$$T(r)=\begin{bmatrix}0 & -1 & 0\\ 1 & 0 & 0\\ 0 & 0 & 1\end{bmatrix},\qquad T(h)=\begin{bmatrix}0 & -1 & 0\\ 0 & 0 & -1\\ 1 & 0 & 0\end{bmatrix}$$

give $\psi(g_4)=1$ and $\psi(g_3)=0$ for $g_3\in C_3$ and $g_4\in C_4$. Now $r^2\in C_5$ and $rh\in C_2$, so we have

$$T(r^2)=\begin{bmatrix}-1 & 0 & 0\\ 0 & -1 & 0\\ 0 & 0 & 1\end{bmatrix},\qquad T(rh)=\begin{bmatrix}0 & 0 & 1\\ 0 & -1 & 0\\ 1 & 0 & 0\end{bmatrix},$$

whence $\psi(g_5)=-1$ and $\psi(g_2)=-1$ for $g_5\in C_5$ and $g_2\in C_2$. Obviously $\psi(1)=3$; hence ψ is tabulated

	C_1	C_2	C_3	C_4	C_5
ψ	3	-1	0	1	-1

and we see that $\psi=\zeta^{(5)}$. The fact that $\langle\psi,\zeta^{(i)}\rangle=0$ for $1\le i\le 4$ can be checked, though it follows from (14.5).

Now we continue the example of A_5 begun at the end of section 14. For each $g\in A_5$, let $T(g)$ be the permutation matrix having 1 as its (i, j)-entry if $g: i\to j$ and having 0 in all other entries. Then T is a representation of A_5, and as in our eight-dimensional representation of S_4 above, if χ is the character associated with T, then for each $g\in A_5$, $\chi(g)$ is just the number of points left fixed by g. Hence we have the character

χ	5	1	2	0	0

using the conjugate classes listed in section 14. Now since we know A_5 has the trivial character $\zeta^{(1)}$, we compute

$$\langle\chi,\zeta^{(1)}\rangle=\tfrac{1}{60}(1\cdot 5+15\cdot 1+20\cdot 2)=1;$$

thus the multiplicity of $\zeta^{(1)}$ in χ is 1. Now let $\psi=\chi-\zeta^{(1)}$. We have

ψ	4	0	1	-1	-1

The only way in which ψ might reduce is as the sum of two characters of degree 2 since $\zeta^{(1)}$ is not a constituent of ψ and there is no other irreducible character of degree 1.

If A_5 has an irreducible character of degree 2, then by (14.13), the sum of the squares of the degrees of the other three irreducible characters is 55. Let $a^2+b^2+c^2=55$ with $a \le b \le c$. If c is 5, 6, or 7, we obtain a^2+b^2 equal to 30, 19, or 6, respectively, none of which is the sum of two squares. On the other hand, if $c \le 4$, then $a^2+b^2+c^2 \le 48$. Hence A_5 can have no irreducible character of degree 2, ψ is irreducible, and (by the argument at the end of section 14) the remaining irreducible characters of A_5 have degree 3, 3, and 5.

We shall be able to complete this example in section 17.

EXERCISES

15.1. For the character χ of D_4 given in the text following (15.3), show that $\langle \chi, \zeta^{(i)} \rangle = 0$ for $1 \le i \le 4$.

15.2. Form a representation U of S_4 by taking e_1, e_2, e_3, e_4 in $\mathbf{R}^4$ as corresponding to the diagonals a, b, c, d of the cube (see figure 9) and by letting $U(g)$ be defined by analogy with T afforded by $\mathbf{R}^8$. Find the values of the character of U, and reduce this character to the sum of its irreducible components.

15.3. In the notation of exercise 15.2, let

$$A = \overline{1234} \qquad C = \overline{1485} \qquad E = \overline{3487}$$
$$B = \overline{1265} \qquad D = \overline{2376} \qquad F = \overline{5678}$$

be the faces of the cube. Form a representation V by analogy with U in exercise 15.2, find its character, and reduce the character to its irreducible components.

15.4. Show that the group

$$H = \langle r, s : r^5 = s^4 = 1,\ rs = sr^2 \rangle$$

has order 20. (*Hint:* Show that $s^i r^j s^m r^n = s^{i+1} r^{2j} s^{m-1} r^n$ for any nonnegative integers i, j, m, n, and thus that the exponent on s^m can be brought down to 0. Then explain why G has at most twenty elements. Finally, show that G has twenty distinct elements of the form $s^i r^j$.)

15.5. Show that $\sigma(a) = r$, $\sigma(b) = s^3$ determines an isomorphism of the group G of exercise 13.9 onto the group H of exercise 15.4. (*Hint:* To prove that σ is a homomorphism, show that

$$\sigma(a)^5 = \sigma(b)^4 = 1 \quad \text{and} \quad \sigma(b)\sigma(a) = \sigma(a)^2\sigma(b).$$

To prove that σ is onto, find $g \in G$ such that $\sigma(g) = s$.)

15.6. Find the five conjugate classes and the character table of the group H of exercise 15.4.

15.7. Show that

$$U(r)=I_3, \qquad U(s)=\begin{bmatrix} 1 & 0 & -1-i \\ -1+i & -i & 1-i \\ 0 & 0 & -i \end{bmatrix}$$

determines a representation of the group H of exercise 15.4. Find its character ψ, and reduce ψ to its irreducible components.

15.8. One of the nonabelian groups of order 16 is isomorphic to $D_4 \times Z_2$. Find its conjugate classes and its character table. (*Hint:* The derived group has order 2.)

15.9. Another nonabelian group of order 16 is the generalized quaternion group

$$Q_4 = \langle a, b : a^8 = 1, b^2 = a^4, ba = a^{-1}b \rangle.$$

Given that $Q_4' = \langle a^2 \rangle$, and that $Q_4/Q_4' \cong V_4$, find the character table for Q_4.

15.10. Still another nonabelian group of order 16 (there are nine in all) is

$$M = \langle x, y : x^8 = y^2 = 1, yx = x^5 y \rangle.$$

Find its conjugate classes and character table. (*Hint:* $C(M) = \langle x^2 \rangle$ and $M' = \langle x^4 \rangle$.)

15.11. Find the character table of D_6, the dihedral group of order 12.

15.12. There are three nonabelian groups of order 18. Two are the dihedral group D_9 and the direct product $S_3 \times Z_3$; the third is

$$G = \langle a, b, c : a^3 = b^3 = c^2 = 1, ba = ab, ca = a^{-1}c, cb = b^{-1}c \rangle.$$

Find the five conjugate classes and the character table of G.

15.13. Find the conjugate classes and the character table of D_9.

15.14. Find the conjugate classes and the character table of $S_3 \times Z_3$.

Part Four
ADDITIONAL TOPICS

This final part is devoted to five topics appealing to a variety of interests. The sections are independent except for a reference in section 17 to one result from section 16, and the reader may select those suiting his interests.

In section 16 we consider characters all of whose values are real numbers, but perhaps the main interest in the section is in two modest indications of how character theory can be used to prove results in abstract group theory.

In section 17 we see how a representation of a subgroup may be "lifted" to a representation of the whole group. This technique is illustrated in the completion of the character table for A_5.

A discussion of space groups as they appear in chemistry forms the central part of section 18; the beginning and concluding parts apply abstract group-theoretic ideas to physical problems. The discussion of the "pure" notion of semidirect products is set in a physical science motivation, which may help the mathematics student to see how group structures occur in nature.

We turn in section 19 to infinite groups occurring in physics and work out in detail the curious connection between a complex matrix group in two dimensions and a real matrix group in three dimensions.

In section 20 we discuss the Burnside Counting Theorem with a simple but illustrative application in chemistry and with a discussion of its relation to characters as presented in section 15. This last section is probably the easiest of the five, and it may prove helpful to read it first in order to have some practice with the material of section 15.

Section 16

Real Characters

Most of the entries in the character tables we found in sections 13–15 were real numbers, in fact, integers. In this section we shall see under what circumstances we may expect the characters of a group to have real values. We shall then use our results to give an illustration of how character theory may be used to prove a theorem of abstract group theory.

Let G be a finite group, T be a representation of G, and χ be the character of T. We define the *conjugate representation* T^* of T by

$$T^*(g) = \overline{T(g)} \quad \text{for each } g \in G,$$

and let χ^* denote the character of T^*. That T^* is a representation follows from (13.16). We shall call χ a *real character* if $\chi = \chi^*$, that is, if $\chi(g) = \chi^*(g)$ for each $g \in G$, which condition is equivalent to $\chi(g) \in \mathbf{R}$.

If C_i is a conjugate class of a finite group G, we define the *inverse* of C_i to be $C_i^* = \{g^{-1} : g \in C_i\}$. The designation C_i^* makes sense in view of (13.19), which shows that for $g \in G$,

$$\chi(g^{-1}) = \overline{\chi(g)} = \chi^*(g).$$

We call C_i a *self-inverse conjugate class* if $C_i = C_i^*$.

Note that C_i^* is indeed a conjugate class. To prove this, we observe that (using the standard symbol $\exists$ for "there exists")

$$\begin{aligned} g, h \in C_i \quad &\text{iff} \quad \exists x \in G : x^{-1}gx = h \\ &\text{iff} \quad \exists x \in G : (x^{-1}gx)^{-1} = h^{-1} \\ &\text{iff} \quad \exists x \in G : x^{-1}g^{-1}x = h^{-1} \\ &\text{iff} \quad g^{-1}, h^{-1} \text{ are conjugate.} \end{aligned}$$

Observe, too, that if C_i consists of elements of order 2 (recall exercise 5.10), then $C_i^* = C_i$ since each element is its own inverse. If the order of the elements in C_i is greater than 2, then C_i may or may not be self-inverse; in section 14 we found conjugate classes of A_4 for which $C_3^* = C_4$, whereas in S_4 all conjugate classes are self-inverse.

(16.1) PROPOSITION. If $\zeta^{(i)}$ is an irreducible character of a finite group G, then the character $\zeta^{(i)*}$ is also irreducible.

PROOF. Let Z be the representation whose character is $\zeta^{(i)}$; then Z is irreducible. Define $T(g) = (Z(g^{-1}))^t$ for each $g \in G$, where the superscript ^{t} denotes (as usual) "transpose." Then for $g, h \in G$, we have

$$\begin{aligned} T(gh) &= (Z((gh)^{-1}))^t = (Z(h^{-1}g^{-1}))^t \\ &= (Z(h^{-1})Z(g^{-1}))^t = (Z(g^{-1}))^t(Z(h^{-1}))^t \\ &= T(g)T(h), \end{aligned}$$

so T is a representation of G. Moreover, if T is reducible, then by Maschke's Theorem (11.11) T is decomposable, and there is a nonsingular matrix A such that

$$A^{-1}T(g)A = \left[\begin{array}{c|c} T_1(g) & \bar{0} \\ \hline \bar{0} & T_2(g) \end{array}\right] \qquad \text{for each } g \in G.$$

But then we observe that

$$A^t(A^{-1})^t = (A^{-1}A)^t = I^t = I, \quad \text{so} \quad (A^{-1})^t = (A^t)^{-1},$$

and from this remark we have, for each $g^{-1} \in G$,

$$\begin{aligned} \left[\begin{array}{c|c} T_1(g)^t & \bar{0} \\ \hline \bar{0} & T_2(g)^t \end{array}\right] &= (A^{-1}T(g)A)^t \\ &= A^t(T(g))^t(A^{-1})^t \\ &= A^tZ(g^{-1})(A^t)^{-1}, \end{aligned}$$

so Z is reducible, a contradiction. Hence T is irreducible. Now if χ is the character of T, then χ is irreducible and

$$\begin{aligned} \chi(g) &= \mathrm{tr}((Z(g^{-1}))^t) = \mathrm{tr}(Z(g^{-1})) \\ &= \overline{\zeta^{(i)}(g)} = \zeta^{(i)*}(g) \end{aligned}$$

for all $g \in G$, as was to be proved.

Since the columns of a character table are mutually orthogonal in the sense of (14.11), the table itself forms a nonsingular matrix. We use this observation in the following results:

(16.2) LEMMA. Let A be a nonsingular n-by-n matrix and let π and ϕ be elements of S_n (that is, permutations). Let A_1

be the matrix whose i^{th} row is the $\pi(i)^{\text{th}}$ row of A, and A_2 be the matrix whose j^{th} column is the $\phi(j)^{\text{th}}$ column of A. If $A_1 = A_2$, then π and ϕ leave fixed the same number of points.

Before proving this lemma, let us examine the situation described. Let $\pi = (13)$, $\phi = (132) \in S_3$, and let

$$A = \begin{bmatrix} 1 & 2 & 3 \\ 4 & 5 & 6 \\ 7 & 8 & 9 \end{bmatrix}.$$

Then the first row of A_1 is the third row of A, and the third row of A_1 is the first row of A. The second row of A_1 is the same as the second row of A, since π does not move 2. Thus

$$A_1 = \begin{bmatrix} 7 & 8 & 9 \\ 4 & 5 & 6 \\ 1 & 2 & 3 \end{bmatrix}.$$

By similar reasoning,

$$A_2 = \begin{bmatrix} 3 & 1 & 2 \\ 6 & 4 & 5 \\ 9 & 7 & 8 \end{bmatrix}.$$

Of course, in this example (which was chosen for visual clarity) $A_1 \neq A_2$, and the hypothesis of (16.2) is not fulfilled, but we can also see that π and ϕ do not fix the same number of points.

PROOF of (16.2). Represent π and ϕ by permutation matrices in the usual way, that is, represent π by B and ϕ by C where

$$Be_i = e_{\pi(i)} \quad \text{and} \quad Ce_i = e_{\phi(i)}$$

for the column vectors $e_1, \ldots, e_n$ forming the natural basis of $\mathbf{R}^n$. The result is that $B = (b_{ij})$ where $b_{ij} = 1$ if $i = \pi(j)$ and $b_{ij} = 0$ otherwise. Using the Kronecker delta (see section 14), we express this observation as $B = (\delta_{\pi(j)j})$; similarly, $C = (\delta_{\phi(j)j})$. Then $A_1 = BA$ and $A_2 = AC$. By hypothesis, $BA = AC$, so $A^{-1}BA = C$. Hence by (13.3) $\operatorname{tr}(B) = \operatorname{tr}(C)$. But by the formation of B and C (observing that $b_{ii} = 1$ if and only if $\pi(i) = i$, and similarly for c_{ii}), the trace of B is the number of points left fixed by π, and similarly for C.

Now we are able to prove the first of our two results on real characters:

(16.3) THEOREM. The number of *real* irreducible characters of a finite group G equals the number of self-inverse conjugate classes of G.

PROOF. Let s be (as usual) the number of distinct conjugate classes of G. By (16.1), the mapping $\zeta^{(i)} \to \zeta^{(i)*}$ is a permutation of the irreducible characters and hence of the integers $1, 2, \ldots, s$; let it be the π of (16.2). By the remarks at the beginning of this section, the mapping $C_i \to C_i^*$ is a permutation of the conjugate classes; let it be ϕ. Consider the character table for G as a matrix A whose (i, j)-entry is $\zeta_j^{(i)}$. If we apply π to A as described in the lemma, we obtain

$$A_1 = (\overline{\zeta_j^{(i)}}),$$

and if we apply ϕ to A as in the lemma, we have

$$A_2 = (\zeta_j^{(i)*}).$$

By (13.19),

$$\begin{aligned} \zeta_j^{(i)*} &= \zeta^{(i)}(x^{-1}) \quad \text{for any} \quad x \in C_j \\ &= \overline{\zeta^{(i)}(x)} \\ &= \overline{\zeta_j^{(i)}}; \end{aligned}$$

hence $A_1 = A_2$. Therefore, by the lemma, π and ϕ leave fixed the same number of points of $\{1, 2, \ldots, s\}$. But π fixes precisely the real characters, and ϕ fixes precisely the self-inverse conjugate classes. This completes the proof.

As a consequence of this theorem, we have our other result on real characters:

(16.4) THEOREM. If $|G|$ is odd, then G has only *one* real irreducible character, namely the trivial character $\zeta^{(1)}$, which has $\zeta^{(1)}(g) = 1$ for all $g \in G$.

PROOF. Let C_i be a conjugate class with $C_i = C_i^*$ and let $x \in C_i$. Then there exists $u \in G$ such that $u^{-1}xu = x^{-1}$. Hence

$$u^{-2}xu^2 = u^{-1}(u^{-1}xu)u = u^{-1}x^{-1}u = (u^{-1}xu)^{-1} = x,$$
$$u^{-3}xu^3 = u^{-1}(u^{-2}xu^2)u = u^{-1}xu = x^{-1},$$

and similarly,

$$u^{-r}xu^r = x \qquad \text{when } r \text{ is even,}$$

$$u^{-r}xu^r = x^{-1} \qquad \text{when } r \text{ is odd.}$$

Now by exercise 3.10, the order of u must divide $|G|$ and so must be odd. If r is the order of u, we have

$$x = 1 \cdot x \cdot 1 = u^{-r}xu^r = x^{-1}.$$

But an element that is its own inverse must have order 1 or 2, and since $|G|$ is odd, G can have no elements of order 2. Hence $x = 1$, and $C_1 = \{1\}$ is the only self-inverse conjugate class. By (16.3), G has only one *real* irreducible character. That character must therefore be the trivial one, which every group has. This completes the proof.

An easy illustration of (16.4) is given by Z_3, which by (12.1) has one real irreducible character and two complex ones. From the representations (12.11) of the noncyclic group of order 9, we see that it too has only one irreducible character. Of course, to *apply* (16.4) rather than to *illustrate* it, one uses it in constructing the irreducible characters of a group whose character table is not already known. The most accessible examples are the smallest nonabelian groups of odd order, which are the two noncyclic groups of order 21:

$$(16.5) \qquad \begin{cases} G = \langle x, y : x^3 = y^7 = 1,\ yx = xy^2 \rangle, \\ H = \langle u, v : u^3 = v^7 = 1,\ vu = uv^4 \rangle. \end{cases}$$

If we construct the character table for G (see exercise 16.6) we know to look for complex values in every irreducible character except $\zeta^{(1)}$.

For the remainder of this section it will be convenient to have the notion of congruence for integers.

(16.6) DEFINITION. Let n be a fixed integer greater than 1. Two integers a and b are *congruent modulo n*, denoted

$$a \equiv b \pmod{n}$$

if n divides the difference $a - b$.

Since n divides $a - b$ in (16.6) precisely when there is an integer k (positive, negative, or zero) such that $a - b = kn$, we observe:

(16.7) PROPOSITION. If a, b, n are integers with $n > 1$, then $a \equiv b \pmod{n}$ if and only if there is an integer k such that $a = kn + b$.

The fact that congruence is an equivalence relation has been left as exercise 16.3.

We now translate the third Sylow theorem (6.8) in terms of congruence as follows:

(16.8) THEOREM (Sylow III, restated). Let $|G|=p^e q$, where p is prime and p does not divide q, and let s_p be the number of distinct Sylow p-subgroups of G. Then $s_p \equiv 1 \pmod{p}$, and s_p divides q.

We shall also need to use the following theorem:

(16.9) THEOREM. If $\zeta^{(i)}$ is an irreducible character of a finite group G, then the degree z_i of $\zeta^{(i)}$ divides $|G|$.

This theorem is not at all easy to prove (see Curtis and Reiner [**4**], (33.7), p. 236, or Scott [**8**], 12.2.27, p. 332).

As a prelude to our main example of an abstract group-theoretic theorem with a character-theoretic proof, we shall give contrasting proofs of another interesting result, one proof in "pure" terms and one using character theory. This discussion is based on one shown to me in 1962 by George Glauberman.

(16.10) PROPOSITION. If p and q are primes with $p \geq q$, and if q does not divide $p-1$, then the only group of order pq is abelian.

PROOF. If $p=q$, then the result follows by (12.15). If $p>q$, we observe from (12.16) that our proof that a group of order pq is abelian will show that such a group is in fact *cyclic*. Now let $|G|=pq$ with $p>q$.

FIRST VERSION. By (16.8) the number s_p of Sylow p-subgroups of G may be written as $1+kp$ for some $k \geq 0$, and s_p divides q. Now if $k \geq 1$, then $1+kp>q$ and so $1+kp$ cannot divide q; hence $k=0$. By (6.9), G has a normal Sylow p-subgroup P. Now the number s_q of Sylow q-subgroups of G may be written as $1+mq$ for some $m \geq 0$, and s_q divides p. But the only divisors of p are 1 and p itself, so s_q equals 1 or p. If $1+mq=p$, then $mq=p-1$, which contradicts the hypothesis that q fails to divide $p-1$. Hence $1+mq=1$, from which we have $m=0$ and $s_q=1$, whence G also has a normal Sylow q-subgroup Q. By (6.10) and (12.3), $G \cong Z_p \times Z_q$ and so is abelian.

SECOND VERSION. Let $z_1, \ldots, z_s$ be the degrees of the irreducible characters of G, listed so that $z_j \leq z_{j+1}$ for $1 \leq j \leq s-1$. By (14.13),

$$|G| = pq = z_1^2 + \ldots + z_s^2. \tag{16.11}$$

If $[G:G'] = r$, then $z_j = 1$ for $1 \leq j \leq r$, and $z_j > 1$ for $r+1 \leq j \leq s$. Let $z_j > 1$; by (16.9), z_j divides pq, and since pq is the product of two primes, we must have z_j equal to one of p, q, pq. Since $p > q$, we know that $p^2 > pq$, and hence by (16.11), z_j cannot be p or pq, both of which make z_j^2 too large. Now (16.11) becomes

$$pq = 1(r) + q^2(s-r). \tag{16.12}$$

Since q divides the other terms of this equation, q must also divide r. But $r = [G:G']$, so r divides $pq = |G|$. Now r must therefore be q or pq. If $r = q$, then (16.12) yields

$$s - r = (pq - q)/q^2 = (p-1)/q,$$

and since $s - r$ is an integer, q divides $p-1$, contradicting the hypothesis. Hence $r = pq$, which means that $G' = \{1\}$, and by (12.20), G is abelian.

My reason for exhibiting two different proofs is to show that group-theoretic results may lend themselves to either type of argument. The first version is probably more "elementary" than the second, but the second is a good illustration of how character-theoretic arguments can have a bearing on abstract group theory.

Now we turn to a striking result incorporating what we have discussed about real characters (see Burnside [**2**], pp. 294–295):

(16.13) THEOREM. Let G be a group of odd order having s conjugate classes. Then $s \equiv |G| \pmod{16}$.

PROOF. By (16.4), G has only one irreducible real character $\zeta^{(1)}$, and by (16.1) the remaining irreducible characters come in conjugate pairs. Hence the number of conjugate classes (or equivalently, of irreducible characters) is odd; we may thus write $s = 2m - 1$ and list the irreducible characters of G as

$$\zeta^{(1)}, \zeta^{(2)}, \zeta^{(2)*}, \ldots, \zeta^{(m)}, \zeta^{(m)*}.$$

Now by (16.9), if

$$z_i = \zeta_1^{(i)} = \zeta_1^{(i)*},$$

then z_i is odd. By (14.13),

$$\begin{aligned} |G| &= 1 + z_2^2 + z_2^2 + \ldots + z_m^2 + z_m^2 \\ &= 1 + \sum_{k=2}^{m} 2z_k^2 \end{aligned}$$

and since z_k is odd we can write $z_k = 2y_k + 1$ for $2 \leq k \leq m$, with each $y_k \geq 0$. Hence

$$\begin{aligned} |G| &= 1 + \sum_{k=2}^{m} 2(2y_k + 1)^2 \\ &= 1 + 2 \sum_{k=2}^{m} (4y_k^2 + 4y_k + 1) \\ &= 1 + 2(m-1) + 2 \sum_{k=2}^{m} 4y_k(y_k + 1) \\ &= 2m - 1 + 8 \sum_{k=2}^{m} y_k(y_k + 1). \end{aligned}$$

Now either y_k or $y_k + 1$ must be even for each k, so each factor $y_k(y_k + 1)$ is even, and there is an integer K such that

$$|G| = 2m - 1 + 16K.$$

But $2m - 1 = s$, so $|G| \equiv s \pmod{16}$, as was to be shown.

Note that although the *statement* of the theorem has nothing to do with representations or characters, the elegant proof given here depends upon many results in character theory. A significant feature of characters is the number of proofs they afford of abstract group-theoretic theorems. Such results may be found in Curtis and Reiner [**4**] and other references.

EXERCISES

16.1. Let

$$A = \begin{bmatrix} 1 & 2 & 3 \\ 4 & 5 & 6 \\ 7 & 8 & 9 \end{bmatrix},$$

$\pi = (13)$ and $\phi = (23)$. Verify by this example that the converse of (16.2) is false.

16.2. Verify that (16.3) holds for the groups whose character tables were found in section 14.

16.3. Show that for integers a, b, c, n, with $n > 1$,

$$a \equiv a \pmod{n};$$

$$\text{if } a \equiv b \pmod{n}, \quad \text{then} \quad b \equiv a \pmod{n};$$

$$\text{if } a \equiv b \pmod{n} \quad \text{and} \quad b \equiv c \pmod{n}, \quad \text{then} \quad a \equiv c \pmod{n}.$$

This shows that congruence is an equivalence relation on integers (for fixed n).

16.4. What does (16.3) say for a finite *abelian* group? (*Hint:* What is a conjugate class in an abelian group?) Check this observation for V_4 and Z_6.

16.5. Prove that if $|G|$ is odd and less than 21, then G is abelian.

16.6. Consider the group G of (16.5). Show first that G has exactly three one-dimensional characters. Then show that G has only two other irreducible characters. Find a conjugate class C_2 by considering the conjugates of y. Then use (16.3) to find a class C_3 and representatives of the other two classes. Finally, complete the character table.

16.7. Let p and q be primes with $p > q$, and assume that q does not divide $(p^2 - 1)$. Prove that a group of order $p^2 q$ must be abelian. If G is abelian, must G be cyclic?

Section 17

Induced Representations and Characters

At (12.18) we used representations of a factor group G/N to find representations of the group G. In this section we shall lift representations of a subgroup H of a group G to representations of G; we shall not even assume in general that H is a normal subgroup.

Let $H \leq G$ and let T be a representation of H afforded either by $\mathbf{R}^m$ or by $\mathbf{C}^m$. Define a function $\dot{T}$ on G by

$$\text{(17.1)} \qquad \begin{cases} \dot{T}(g) = T(g) & \text{if } g \in H, \\ \dot{T}(g) = 0_m & \text{if } g \notin H, \end{cases}$$

where m is the degree of T and 0_m denotes the m-by-m matrix consisting entirely of zeros. Note that in general $\dot{T}$ will *not* be a representation; it is merely an intermediate function. Now let $n = [G:H]$ and write

$$\text{(17.2)} \qquad G = H \cup g_2H \cup g_3H \cup \ldots \cup g_nH.$$

Here we have taken the identity 1 of G as the representative of the first coset; for convenience we shall regularly take $g_1 = 1$ and $H = g_1H$. Now for $g \in G$, we define

$$\text{(17.3)} \qquad T^G(g) = (a_{ij}) = (\dot{T}(g_i^{-1}gg_j)).$$

If the degree of T is m, then each T^G is an mn-by-mn matrix consisting of n^2 submatrices, each m-by-m, and each submatrix equal either to some $T(h)$ or to 0_m.

For example, let

$$\text{(17.4)} \qquad \begin{cases} G = D_4 = \langle x, y : x^4 = y^2 = 1,\ yx = x^{-1}y\rangle \\ \text{and } H = \langle x \rangle, \qquad G = H \cup yH. \end{cases}$$

Let $T(x) = [i]$; then

$$\dot{T}(1) = [1], \qquad \dot{T}(x) = [i], \qquad \dot{T}(x^2) = [-1], \qquad \dot{T}(x^3) = [-i],$$
$$\dot{T}(y) = \dot{T}(xy) = \dot{T}(x^2y) = \dot{T}(x^3y) = [0].$$

Now since $g_1 = 1$ and $g_2 = y$, we have

$$1x1 = x, \qquad 1xy = xy, \qquad yx1 = x^3y, \qquad yxy = x^3,$$

and hence by (17.3),

$$T^G(x) = \begin{bmatrix} i & 0 \\ 0 & -i \end{bmatrix}.$$

Similarly, $1\,y\,1 = y$, $1\,yy = 1$, $yy\,1 = 1$, $yyy = y$; hence

$$T^G(y) = \begin{bmatrix} 0 & 1 \\ 1 & 0 \end{bmatrix}.$$

If χ is the character associated with T^G, then

$$\chi(x) = 0, \quad \chi(y) = 0, \quad \chi(x^2) = -2, \quad \chi(xy) = 0, \quad \text{and} \quad \chi(1) = 2.$$

Hence $\chi = \chi_5$ from the character table at the end of section 13.

For another example, take G and H as in (17.4) and consider

$$T(x) = \begin{bmatrix} 0 & -1 \\ 1 & 0 \end{bmatrix};$$

then the powers of $T(x)$ give

$$T(x^2) = \begin{bmatrix} -1 & 0 \\ 0 & -1 \end{bmatrix} \quad \text{and} \quad T(x^3) = \begin{bmatrix} 0 & 1 \\ -1 & 0 \end{bmatrix}.$$

From (17.3) we can then find

$$T^G(x) = \begin{bmatrix} 0 & -1 & 0 & 0 \\ 1 & 0 & 0 & 0 \\ 0 & 0 & 0 & 1 \\ 0 & 0 & -1 & 0 \end{bmatrix}, \qquad T^G(y) = \begin{bmatrix} 0 & 0 & 1 & 0 \\ 0 & 0 & 0 & 1 \\ 1 & 0 & 0 & 0 \\ 0 & 1 & 0 & 0 \end{bmatrix}.$$

If χ is the character of T^G, we have $\chi(x) = 0$, $\chi(y) = 0$, $\chi(x^2) = -4$, $\chi(xy) = 0$, $\chi(1) = 4$; thus $\chi = 2\chi_5$, also from the character table at the end of section 13.

We call the representation T^G an *induced representation*, or more specifically, the representation of G induced by the representation T of H. If χ is the character of H, we shall denote by χ^G the character of T^G and call χ^G an *induced character.*

We can form χ^G from χ by a construction analogous to that for T^G. Let

$$\text{(17.5)} \qquad \begin{cases} \dot{\chi}(g) = \chi(g) & \text{if } g \in H, \\ \dot{\chi}(g) = 0 & \text{if } g \notin H. \end{cases}$$

Then let G be written as in (17.2), and let

$$\chi^G(g) = \sum_{i=1}^{n} \dot{\chi}(g_i^{-1}gg_i). \tag{17.6}$$

Clearly, (17.6) gives the same character χ^G as that obtained from the induced representation in (17.3).

Now if $g \in G$ and $h \in H$, then $h^{-1}gh$ is a conjugate of g and so $\dot{\chi}(h^{-1}gh) = \dot{\chi}(g)$ by (13.5). (Note that $h^{-1}gh \in H$ if and only if $g \in H$.) Hence for each i and for each $h \in H$,

$$\dot{\chi}((g_ih)^{-1}g(g_ih)) = \dot{\chi}(g_i^{-1}gg_i);$$

therefore, we have

$$\chi^G(g) = \frac{1}{|H|} \sum_{x \in G} \dot{\chi}(x^{-1}gx). \tag{17.7}$$

For an application of (17.7), let $H \trianglelefteq G$ and let χ be a character of H. Now for any $x \in G$, conjugation by x is an automorphism of G, and since conjugation permutes the elements of H and is a 1:1 function, it follows that

$$x^{-1}gx \notin H \quad \text{whenever} \quad g \notin H.$$

Hence $\dot{\chi}(x^{-1}gx) = 0$ whenever $g \notin H$, and so $\chi^G(g) = 0$ when $g \notin H$. This observation greatly reduces the labor of computing χ^G when $H \trianglelefteq G$.

An important consequence of (17.7) is:

(17.8) PROPOSITION. Let χ^G be an induced character of G. Then the values $\chi^G(g)$ for $g \in G$ are independent of the set of coset representatives $g_1, \ldots, g_n$ used in (17.2).

As a result of (17.8), we may properly speak of *the* character χ^G induced from the character χ of H.

We may now observe what occurs if $H \leq G$ and if T is the trivial representation of degree 1 on H, with associated character $\zeta^{(1)}$. The submatrices appearing in T^G are single entries, and the rest of each $T^G(g)$ consists of zeros; thus T^G is a representation of G by permutation matrices with the degree of T^G equal to $[G:H]$. Now the coset g_iH (with $1 \leq i \leq n$ as in (17.2)) is contained in the normalizer N_GH if and only if $g_i \in N_GH$. Moreover, a given $T^G(g)$ has a 1 on its main diagonal in the (i, i)-position if and only if $g_i^{-1}gg_i \in H$. In particular, for $h \in H$, the matrix $T^G(h)$ has a 1 in its (i, i)-position if and only if g_i carries h under conjugation into some element of H, that is, if and only if $g_i^{-1}hg_i \in H$. If $g_i \in N_G H$, it will certainly do so, though in fact g_i may normalize a single element h even if g_i fails to normalize all of H (for example, if $h=1$). Thus we see that for $h \in H$, if χ is the character of T^G (for the given T), then $\chi(h) \geq [N_G H : H]$. If $H \trianglelefteq G$, then

$$[N_G H : H] = [G : H] = \text{degree}\ (T^G),$$

so T^G is the trivial representation of degree $[G:H]$ *on the elements of* H, though not necessarily on all of the elements of G. For G and H as at (17.4), we obtain

$$T^G(x) = \begin{bmatrix} 1 & 0 \\ 0 & 1 \end{bmatrix}, \qquad T^G(y) = \begin{bmatrix} 0 & 1 \\ 1 & 0 \end{bmatrix};$$

hence $\chi(x) = \chi(x^2) = \chi(1) = 2$ and $\chi(y) = \chi(xy) = 0$, from which we have $\chi = \chi_1 + \chi_3$ from the character table for D_4.

We have previously spoken of the restriction of the domain of a representation to a subgroup of the original group, for example in (12.21). A character may likewise have its domain restricted. If G is a group, T a representation of G, χ any character of G, and H a subgroup of G, we shall denote by $T|_H$ and $\chi|_H$ the restrictions of T and χ to H.

Recall now the inner product of (14.1) for functions from a group G into the complex numbers $\mathbf{C}$. We can form the inner product of a character of some subgroup H of G with the restriction of a character of G to H. Specifically, let $H \leq G$, let χ be a character of G, and let μ be a character of H. Then it is possible to form $\langle \mu, \chi|_H \rangle$. Similarly, since a character μ^G can be induced from μ, we can form $\langle \mu^G, \chi \rangle$. Although the former product is taken over the subgroup H and the latter over the group G, we can still compare the numerical (complex) values, and we have the striking result:

(17.9) THEOREM. If $H \leq G$, if χ is a character of G, and if μ is a character of H, then

$$\langle \mu, \chi|_H \rangle = \langle \mu^G, \chi \rangle.$$

PROOF. $\langle \mu^G, \chi \rangle = \dfrac{1}{|G|} \sum_{g \in G} \mu^G(g)\, \overline{\chi(g)}$

$$= \frac{1}{|G|}\frac{1}{|H|} \sum_{g \in G} \sum_{x \in G} \dot{\mu}(x^{-1}gx)\, \overline{\chi(g)} \qquad \text{by (17.7)}$$

$$= \frac{1}{|G|\,|H|} \sum_{g \in G} \sum_{x \in G} \dot{\mu}(x^{-1}gx)\, \overline{\chi(x^{-1}gx)} \qquad \text{by (13.5).}$$

Now since conjugation by x is an automorphism of G, a summation over $g \in G$ is precisely the same as a summation over all $x^{-1}gx \in G$. Set $y = x^{-1}gx$; then

$$\langle \mu^G, \chi \rangle = \frac{1}{|G|\,|H|} \sum_{y \in G} \sum_{x \in G} \dot{\mu}(y)\, \overline{\chi(y)}.$$

But now x does not appear in the summands, and hence the effect of the summation over $x \in G$ is merely to add up the same term $|G|$ number of times. Hence

$$\begin{aligned} \langle \mu^G, \chi \rangle &= \frac{1}{|H|} \sum_{y \in G} \dot{\mu}(y)\, \overline{\chi(y)} \\ &= \frac{1}{|H|} \sum_{y \in H} \mu(y)\, \overline{\chi(y)} \\ &= \langle \mu, \chi|_H \rangle \end{aligned}$$

since $\dot{\mu}(y) = 0$ when $y \notin H$. This completes the proof.

As a consequence of (17.9) we have the following famous result:

(17.10) FROBENIUS RECIPROCITY THEOREM. Let $H \leq G$, χ be an irreducible character of G, and μ an irreducible character of H. Then the multiplicity of χ in μ^G equals the multiplicity of μ in $\chi|_H$.

PROOF. In the terminology of section 15, what we have to prove is that $\langle \mu^G, \chi \rangle = \langle \chi|_H, \mu \rangle$. Now $\langle \chi|_H, \mu \rangle$ is a nonnegative (real) integer, so

$$\langle \chi|_H, \mu \rangle = \overline{\langle \chi|_H, \mu \rangle} = \langle \mu, \chi|_H \rangle.$$

The result follows by (17.9).

For an illustration, consider the irreducible characters of A_4 and of S_4 that were found in section 14. Let $G = S_4$, $H = A_4$; let $\chi = \zeta^{(4)}$ from the character table for S_4, and $\mu = \zeta^{(4)}$ from the character table for A_4. Then

$\chi\|_H$	3	−1	0	0

and clearly $\chi|_H = \mu$, so $\langle \chi|_H, \mu \rangle = 1$. Hence $\langle \mu^G, \chi \rangle = 1$. Now for the same G, H, and μ, let $\chi = \zeta^{(5)}$ for S_4. Then

$\chi\|_H$	3	−1	0	0

and again $\langle \mu^G, \chi \rangle = 1$. Thus two distinct irreducible characters for S_4 give the same irreducible character of A_4 upon restriction to A_4, and by (17.10), *both* of these irreducible characters are constituents of the character for S_4 induced from the one irreducible character of A_4. Now note that

$$\text{degree }(\mu^G) = (\deg(\mu))[S_4:A_4] = 6,$$

and the two constituents we have found for μ^G each have degree 3. Hence for the G, H, and μ above and for the irreducible characters $\zeta^{(4)}$ and $\zeta^{(5)}$ of S_4, we have

$$\mu^G = \zeta^{(4)} + \zeta^{(5)}.$$

The power of the Frobenius Reciprocity Theorem is seen in that we were able to find the values of μ^G and to resolve μ^G into its irreducible components without computing the induced representation whose character is μ^G.

As a further application of (17.10), we prove:

(17.11) THEOREM. Let $H \leq G$, $[G:H] = m$, and ζ be an irreducible character of G. Then there is an irreducible character χ of H such that $m \cdot \text{degree }(\chi) \geq \text{degree }(\zeta)$.

PROOF. Let χ be an irreducible character of H such that $\langle \zeta|_H, \chi \rangle \neq 0$. Then by (17.10), $\langle \chi^G, \zeta \rangle \neq 0$. Hence

$$\text{degree }(\zeta) \leq \text{degree }(\chi^G) = m \cdot \text{degree }(\chi).$$

(17.12) COROLLARY. If G has an abelian subgroup H of index m, then the degree of every irreducible character of G is less than or equal to m.

PROOF. Since H is abelian, the degree of χ in theorem (17.11) must be 1.

Corollary (17.12) can be used to construct a character table in order to determine the maximum possible degree for an irreducible character. The larger the abelian subgroup H that can be found, the smaller its index, and hence the smaller the degree of an irreducible representation must be.

Conversely, when we study the subgroup structure of a group whose irreducible characters are known, the maximum degree of an irreducible character gives a lower bound for the index of an abelian subgroup and hence an upper bound for the order of abelian subgroups of the given group.

To conclude this section, we shall complete the character table for A_5, which we started in section 14 and worked on in section 15. Let $G = A_5$ and $H = A_4$, and consider $\chi = \zeta^{(2)}$ from the table for A_4 (section 14). We can write

$$G = H \cup (152)H \cup (253)H \cup (245)H \cup (354)H$$

and compute

$$\begin{aligned}
\chi^G(1) &= \chi(1) \cdot 5 = 5, \\
\chi^G((12)(35)) &= \dot{\chi}((12)(35)) + \dot{\chi}((125)(12)(35)(152)) \\
&\quad + \dot{\chi}((235)(12)(35)(253)) \\
&\quad + \dot{\chi}((254)(12)(35)(245)) \\
&\quad + \dot{\chi}((345)(12)(35)(354)) \\
&= 0 + \dot{\chi}((15)(23)) + \dot{\chi}((15)(23)) \\
&\quad + \dot{\chi}((14)(23)) + \dot{\chi}((12)(45)) \\
&= 0+0+0+1+0 \\
&= 1,
\end{aligned}$$

and similarly,

$$\begin{aligned}
\chi^G((123)) &= \zeta + \dot{\chi}((135)) + \dot{\chi}((152)) + \dot{\chi}((143)) + \dot{\chi}((125)) \\
&= \zeta + \zeta^2 \\
&= -1, \\
\chi^G((12345)) &= 0 + \dot{\chi}((13425)) + \dot{\chi}((15243)) \\
&\quad + \dot{\chi}((14352)) + \dot{\chi}((12534)) \\
&= 0, \\
\chi^G((12354)) &= 0.
\end{aligned}$$

Now $\langle \chi^G, \zeta^{(1)} \rangle = (5+15-20)/60 = 0$; hence $\zeta^{(1)}$ is *not* a constituent of χ^G. Since the only other irreducible characters of A_5 have degree 3, 3, 4, and 5, it follows that χ^G can only be the irreducible character $\zeta^{(5)}$ of degree 5.

We summarize the information we have obtained thus far in the following table, indicating the unknown entries by letters:

	C_1	C_2	C_3	C_4	C_5
h_i	1	15	20	12	12
$\zeta^{(1)}$	1	1	1	1	1
$\zeta^{(2)}$	3	a	c	p	r
$\zeta^{(3)}$	3	b	d	q	s
$\zeta^{(4)}$	4	0	1	-1	-1
$\zeta^{(5)}$	5	1	-1	0	0

Application of (14.11) to the second and first columns gives $1+3a+3b+5=0$, whence $a+b=-2$. The same procedure with the remaining columns and the first gives

$$c+d=0, \qquad p+q=1, \qquad r+s=1.$$

Now since C_2 and C_3 consist of all elements of order 2 and 3, respectively, they are self-inverse conjugate classes. We took C_4 to be the class containing (12345); since

$$(14)(23)(12345)(14)(23)=(15432),$$

the inverse of (12345) is in C_4. Now C_4^* is a conjugate class (from section 16) having the element (15432) in common with C_4; hence $C_4=C_4^*$ and as a result $C_5^*=C_5$ as well. By (16.3), all of the characters of A_5 are real; hence all of the unknowns in the table above are real numbers.

If we apply (14.12) to the second column, we have

$$1+a^2+b^2+1=4,$$

which together with $a+b=-2$ reduces to $a=-1$, $b=-1$. Application of (14.12) to column 3 gives $c^2+d^2=0$, whose unique real solution is $c=d=0$. Note that in these computations we have used the fact that the unknowns are real to write a^2 for $a\bar{a}$, etc.

From the second and fourth rows we obtain $r=1-p$ by (14.7). Finally, from the fourth column we have by (14.12)

$$1+p^2+(1-p)^2+1=60/12,$$

which may be solved by the quadratic formula for

$$p=\frac{1\pm\sqrt{5}}{2}.$$

Now $1-p$ is the same as p except for the opposite choice of sign, so we may take + in p and − in $1-p$. This completes the following character table for A_5:

	C_1	C_2	C_3	C_4	C_5
$\zeta^{(1)}$	1	1	1	1	1
$\zeta^{(2)}$	3	−1	0	$(1+\sqrt{5})/2$	$(1-\sqrt{5})/2$
$\zeta^{(3)}$	3	−1	0	$(1-\sqrt{5})/2$	$(1+\sqrt{5})/2$
$\zeta^{(4)}$	4	0	1	−1	−1
$\zeta^{(5)}$	5	1	−1	0	0

EXERCISES

17.1. The group S_4 has a dihedral normal subgroup P generated by the elements $(1423)=x$ and $(12)=y$. Then

$$T(x)=\begin{bmatrix} 0 & 1 \\ -1 & 0 \end{bmatrix}, \qquad T(y)=\begin{bmatrix} -1 & 0 \\ 0 & 1 \end{bmatrix}$$

determines a representation of P. Let χ be the character of T. Find χ^G, and reduce χ^G to its irreducible components.

17.2. The group $G=S_4$ has a normal subgroup N of order 4 generated by (12)(34) and (13)(24). Let χ be a character of degree 1 on N given by $\chi((12)(34))=1$, $\chi((13)(24))=-1$. Let ψ be the character for N associated with the representation given by

$$U((12)(34))=\begin{bmatrix} -1 & 0 \\ 0 & 1 \end{bmatrix}, \qquad U((13)(24))=\begin{bmatrix} 1 & 0 \\ 0 & -1 \end{bmatrix}.$$

Find χ^G and ψ^G, and reduce each to its irreducible components.

17.3. Let $G=A_5$, $H=A_4$, and μ be the irreducible character $\zeta^{(4)}$ of A_4. Use the Frobenius Reciprocity Theorem to reduce μ^G to its irreducible components. Do *not* compute μ^G by (17.7).

17.4. Prove the following:

(17.13) THEOREM (Transitivity of Induction). Let G be a finite group, $K \leq H \leq G$, and χ be a character of K. Then $(\chi^H)^G = \chi^G$.

17.5. For matrices $A=(a_{ij})$ and $B=(b_{ij})$ of degree n and m, respectively, we define the *tensor product* of A and B as

$$A \otimes B = \begin{bmatrix} a_{11}B & a_{12}B & \dots & a_{1n}B \\ a_{21}B & a_{22}B & \dots & a_{2n}B \\ & \dots & & \dots \\ a_{n1}B & a_{n2}B & \dots & a_{nn}B \end{bmatrix}.$$

If T and U are representations of a group G, we define the *tensor product* $T \otimes U$ by

$$(T \otimes U)(g) = T(g) \otimes U(g)$$

for each $g \in G$. For the representation T of S_4 given by (9.4) and (9.5), find $T \otimes T$ and its associated character χ, and reduce χ to its irreducible components.

17.6. Let T and U be representations of a group G with associated characters χ and μ, respectively. Prove that the character of $T \otimes U$

is $\chi\mu$, that is,

$$(\chi\mu)(g) = \chi(g)\mu(g)$$

for each $g \in G$.

17.7. Find the character table of S_5.

17.8. Prove that T^G as defined in (17.3) is a representation of G. (*Hint:* Show that for fixed g_i and g, there is exactly one g_j such that $g_i^{-1} g g_j \in H$.)

Section 18

Space Groups and Semi-direct Products

In section 3 we found two symmetry groups for the cube, one of order 24 that is usual in mathematical treatments and one of order 48 that occurs in chemical applications. In exercise 3.7 we wrote the larger group G^* as the union of the smaller group G and a coset Gc whose representative c was a reflection in a plane parallel to two opposite faces of the cube. We shall begin this section by proving a theorem that determines the character table of G^* in terms of that of G, which we found in section 14 (in view of the isomorphism between G and S_4 found at (15.4)). The proof of the theorem will be organized by means of the following three lemmas.

(18.1) LEMMA. Let $H \leq G$, $[G:H]=2$, a and b be elements of H, and $z \in C(G)$ with $z \notin H$. Then a and b are conjugate in H if and only if az and bz are conjugate in G.

PROOF. If a and b are conjugate in H, then there exists $h \in H$ such that $h^{-1}ah = b$. Then $h^{-1}ahz = bz$, and since $z \in C(G)$, we have $h^{-1}azh = bz$. Conversely, if az and bz are conjugate in G, then there exists $g \in G$ such that $g^{-1}azg = bz$. Since $z \in C(G)$, we have $g^{-1}agz = bz$, and by cancellation, $g^{-1}ag = b$. Now since $[G:H]=2$ and $z \notin H$, we have $G = H \cup Hz$, so either $g \in H$, in which event a and b are conjugate in H, or else $g = hz$ for some $h \in H$. In the latter event, $g^{-1}ag = (hz)^{-1}a(hz) = h^{-1}ah$ since $z \in C(G)$.

(18.2) LEMMA. Let $H \leq G$, $[G:H]=2$, $z \in C(G)$, $z \notin H$, and $C_1, \ldots, C_s$ be the distinct conjugate classes of H. Then the distinct conjugate classes of G are

$$C_1, \ldots, C_s, C_1z, \ldots, C_sz,$$

where $C_iz = \{hz : h \in C_i\}$ for $1 \leq i \leq s$.

PROOF. We show first that the classes C_i of H remain distinct conjugate classes when considered as subsets of G. Let $a \in C_i$ and $b \in C_j$, and suppose there exists $h \in H$ such that

$$b = (hz)^{-1}a(hz).$$

Then $b = h^{-1}ah$ since $z \in C(G)$, and hence $i = j$; thus $C_1, \ldots, C_s$ are contained in s distinct conjugate classes of G. By (18.1) the subsets $C_1 z, \ldots, C_s z$ are also contained in s distinct conjugate classes of G, so the only question remaining is whether some C_i and some $C_j z$ might be in the same class. But $H \trianglelefteq G$ by (7.12), and $C_i \subseteq H$, so no element of C_i can be conjugated out of C_i by an element of G; hence $C_i \cap C_j z = \varnothing$ for every i and j. Thus the $2s$ subsets listed are all distinct conjugate classes, and since $G = H \cup Hz$, they account for all of the elements of G.

(18.3) LEMMA. Let $H \leq G$, $[G:H] = 2$, $z \in C(G)$, $z \notin H$; for $1 \leq i \leq s$, let $\zeta^{(i)}$ be an irreducible character of H, and T_i be the representation of H affording $\zeta^{(i)}$. Define $\hat{T}_i$ and $\tilde{T}_i$ on G by

$$\hat{T}_i(hz^k) = T_i(h) \qquad \text{for } k = 0, 1;$$

$$\tilde{T}_i(hz^k) = (-1)^k T_i(h) \qquad \text{for } k = 0, 1.$$

Then each $\hat{T}_i$ and each $\tilde{T}_i$ is an irreducible representation of G. Moreover, if $\hat{\chi}_i$ and $\tilde{\chi}_i$ are the respective characters, then

$$\hat{\chi}_i(hz^k) = \chi(h) \qquad \text{for } k = 0, 1;$$

$$\tilde{\chi}_i(hz^k) = (-1)^k \chi_i(h) \qquad \text{for } k = 0, 1.$$

PROOF. Since for k and j each equal to 0 or 1, and for $h_1, h_2 \in H$,

$$\begin{aligned} \hat{T}_i((h_1 z^k)(h_2 z^j)) &= \hat{T}_i(h_1 h_2 z^{k+j}) \\ &= T_i(h_1 h_2) = T_i(h_1) T_i(h_2) \\ &= \hat{T}_i(h_1 z^k) \hat{T}_i(h_2 z^j), \end{aligned}$$

each $\hat{T}_i$ is a representation of G, and similarly for $\tilde{T}_i$. To verify that $\hat{T}_i$ is well defined, we need only observe that if $h_1 z^k = h_2 z^j$, then $h_2^{-1} h_1 = z^{j-k}$, and since $z \notin H$, $h_1 = h_2$ and $k = j$. If $\hat{T}_i$ is reducible, then the nonsingular matrix A that reduces $\hat{T}_i$ will also reduce T_i; hence $\hat{T}_i$ is irreducible, and similarly for $\tilde{T}_i$. The computation of the characters is obvious.

Now by (18.2) and (18.3) we have proved:

(18.4) THEOREM. If $H \leq G$, $[G:H] = 2$, $z \in C(G)$, and $z \notin H$, then the character table for G is

$C_1 \ldots C_s$	$C_1 z \ldots C_s z$
$\zeta_j^{(i)}$	$\zeta_j^{(i)}$
$\zeta_j^{(i)}$	$-\zeta_j^{(i)}$

where $(\zeta_j^{(i)})_{1 \leq i, j \leq s}$ is the character table for H.

To apply (18.4) to the two groups G^* and G of the cube, we need an element $z \in C(G^*)$ such that $z \notin G$. The element c of exercise 3.7 is not in the center of G^*, but we may set

$$\begin{aligned} z = r^2c &= (13)(24)(57)(68)(15)(26)(37)(48) \\ &= (17)(28)(35)(46). \end{aligned}$$

Since $z \in Gc$, we know that $z \notin G$. Moreover, we remarked in section 9 that $G = \langle r, x \rangle$, where $r = (1234)(5678)$ and $x = (245)(386)$. (See exercise 3.4 and figure 4.) Since

$$\begin{aligned} rz &= (1234)(5678)(17)(28)(35)(46) \\ &= (1836)(2547) = (17)(28)(35)(46)(1234)(5678) \\ &= zr \end{aligned}$$

and

$$\begin{aligned} xz &= (245)(386)(17)(28)(35)(46) \\ &= (17)(265843) = zx, \end{aligned}$$

z commutes with the generators of G and hence with all of the elements of G. Moreover, since $z \notin G$, we have $G^* = G \cup Gz$ and clearly z commutes with every element of Gz; hence $z \in C(G^*)$. Therefore, G^* has ten irreducible characters, and the table for G^* may be obtained from the character table of G determined in section 14 by means of (18.4).

In view of the correspondence worked out at (15.5), we may remark that the effect of z above is to reflect each vertex of the cube through the center of the cube; in applications z is called an *inversion*, with the center of the cube as an *inversion center.* Note that z is not a *rigid* motion in the sense of one that can be performed on a geometric cube without disassembling the object; in fact, the effect of z upon a hollow cube is to turn the figure inside out. On the other hand, if one thinks of a cube as consisting only of its eight vertices (as in the symmetry of crystalline cesium chloride), the motion z has physical meaning.

A similar analysis for the square bipyramid has been left as exercise 18.1.

The idea of regarding a cube as consisting only of its vertices, as one does in interpreting the vertices as atoms in a crystal, leads us to the concepts of point and space groups used in chemistry. We imagine an infinite (or potentially infinite) lattice of points in the plane or in space, such as the points (x, y) in $\mathbf{R}^2$ or (x, y, z) in

$\mathbf{R}^3$ with *integer* coordinates. These examples give square and cubical configurations, respectively, but of course others are possible.

A *space group* may be defined as that collection of geometrical transformations that leaves a lattice of points invariant. These motions consist of rotations (about the origin), reflections, and translations. A point X of a lattice is moved to another point of the lattice by a matrix operation (that is, a linear transformation specifying a rotation or a reflection) followed by a translation:

$$X' = RX + t. \tag{18.5}$$

The matrix operation R is called a *point operation.* We may represent the transformation (18.5) in the Seitz notation $\{\hat{R} \mid \vec{t}\}$ used in much of the chemical literature, or more briefly by (R, t). The operation in the space group is given by

$$(R, t)(S, u) = (RS, Ru + t); \tag{18.6}$$

here RS is again a point operation, and $Ru + t$ is a translation. Note that Ru is indeed a translation; it moves the lattice by the length of u in the direction of the vector to which the rotation R carries the translation vector u. For example, if u is a unit vector in the direction of the positive X-axis and R is a rotation through 90° carrying the positive X-axis to the positive Y-axis, then Ru is a translation of one unit in the positive Y-direction.

The identity of a space group will be represented by $(E, 0)$, where E is the identity transformation and 0 is a translation through distance 0. To check the associative property, we calculate

$$\begin{aligned}[(Q, t)(R, u)](S, v) &= (QR, Qu + t)(S, v) \\ &= (QRS, QRv + Qu + t) \\ &= (Q, t)(RS, Rv + u) \\ &= (Q, t)[(R, u)(S, v)].\end{aligned}$$

The inverse of (R, t) is $(R^{-1}, -R^{-1}t)$ since

$$(R, t)(R^{-1}, -R^{-1}t) = (RR^{-1}, R(-R^{-1})t + t) = (E, 0)$$

and

$$(R^{-1}, -R^{-1}t)(R, t) = (R^{-1}R, R^{-1}t - R^{-1}t) = (E, 0).$$

Thus we have (assuming closure) a space group G, which in fact is a group acting on the set of lattice points in the sense of (2.4).

We choose a set of *basic primitive translations,* one for each dimension of the $\mathbf{R}^n$ under consideration. A linear combination of basic primitive translations using integer coefficients is called a *primitive translation.* Because the point operations permute the points of the lattice, Ru will be a primitive translation whenever R is a point operation and u a primitive translation.

To simplify our discussion, let us restrict ourselves to lattices in the plane $\mathbf{R}^2$. We begin with a lattice L consisting of all points (x, y) with integer coefficients. The basic primitive translations t_1 and t_2 will be movements of the entire lattice one unit to the right and one unit up, respectively. Thus, for example, t_1+t_2 will move the lattice $\sqrt{2}$ units up the line $x=y$. For point operations we consider a rotation C_4 through 90° counterclockwise around the (fixed) origin. (Here C_n is the usual symbol in chemical applications for a rotation through $2\pi/n$; thus $C_4^2=C_2$ and $C_4^4=E$.)

The operation (C_4, t_1) rotates the lattice through 90° and then shifts it one unit to the right. The elements (C_4, t_1) and $(C_4, t_1)^2$ may be seen in figure 10 for nine points of L. Note that by (18.6) we have

$$(C_4, t_1)(C_4, t_1)=(C_4^2, C_4t_1+t_1),$$

where $C_4t_1=t_2$, as described following (18.6). The reader should verify geometrically that $(C_4, t_1)^4=(E, 0)$. Further examples are given in exercise 18.3.

Now the set of primitive translations forms a subgroup of G, as is clear from the computation

$$(E, t)(E, u)=(E, Eu+t)=(E, u+t)$$

for closure and the existence of an inverse $(E, -t)$ for (E, t) (see (3.1)). Moreover, if R is a point operation and if t and u are primitive translations, then

$$\begin{aligned}(R, t)^{-1}(E, u)(R, t) &= (R^{-1}, -R^{-1}t)(E, u)(R, t)\\ &= (R^{-1}, R^{-1}u-R^{-1}t)(R, t)\\ &= (E, R^{-1}t+R^{-1}u-R^{-1}t)\\ &= (E, R^{-1}u);\end{aligned}$$

hence the set T of all primitive translations forms a normal subgroup of G.

The point operations of a space group will form a group P, called the *point group*, which is isomorphic to G/T, but which in general *need not be a subgroup* of G. Even if P is a subgroup of G, so that we can write typical elements in the form $(R, 0)$, we still

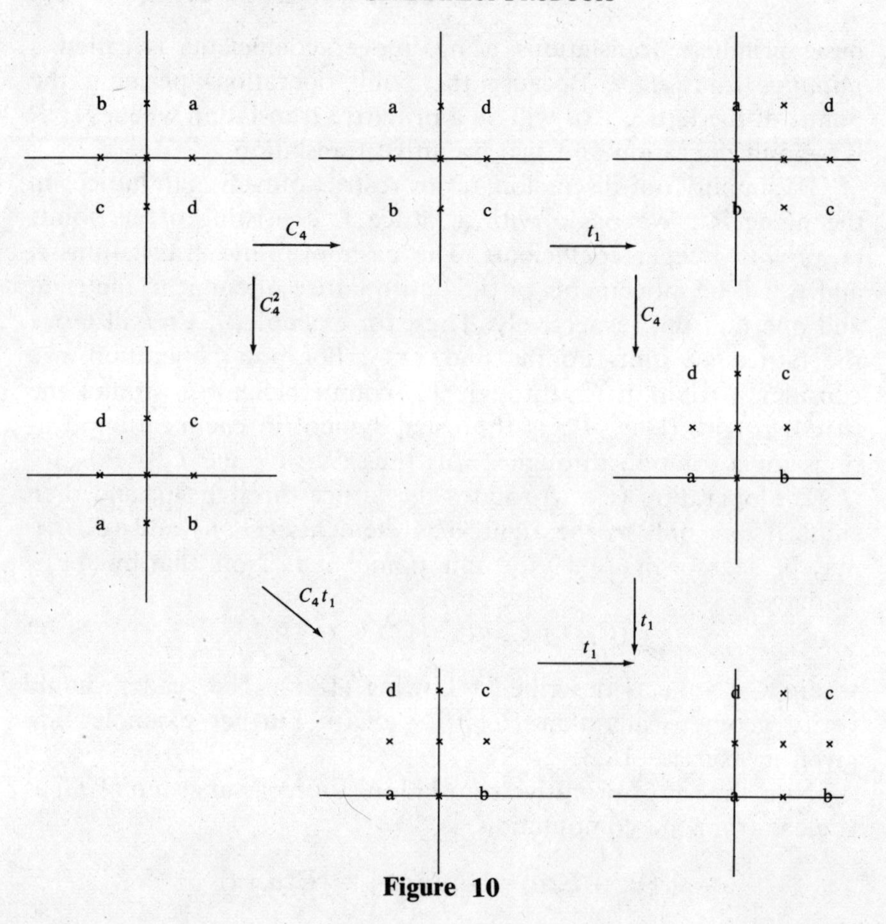

Figure 10

have

$$
\begin{aligned}
(S, t)^{-1}(R, 0)(S, t) &= (S^{-1}, -S^{-1}t)(R, 0)(S, t) \\
&= (S^{-1}R, -S^{-1}t)(S, t) \\
&= (S^{-1}RS, S^{-1}Rt - S^{-1}t),
\end{aligned}
$$

which will not in general have $S^{-1}Rt - S^{-1}t = 0$; thus P is not normal in G, and in particular, G is not the direct product of P and T. A space group in which the point group is a (nonnormal) subgroup is called *symmorphic;* we shall examine a mathematical characterization of symmorphic space groups later in this section.

Next let us find the irreducible representations of a group T of primitive translations. To avoid the problems of infinite space groups and to avoid surface problems (lattice sites at a surface of a crystal are obviously in a different environment from those in the bulk), we resort to the use of the *cyclic boundary condition:* once one reaches a boundary of a crystal (or plane figure), he immediately moves to a corresponding location on the opposite boundary. For each of the basic primitive translations t_1, t_2 in $\mathbf{R}^2$ or t_1, t_2, t_3 in $\mathbf{R}^3$, we assume that there is an integer N_j such that

$$(E, t_j)^{N_j} = (E, 0). \tag{18.7}$$

Equivalently, $(E, N_j t_j) = (E, 0)$ for each j. Then the dimensions of the figure or crystal are

$$N_1\,|t_1|\text{-by-}N_2\,|t_2|(\text{-by-}N_3\,|t_3|).$$

The group T of primitive translations is obviously abelian, and thus is the direct product of the $\langle(E, t_i)\rangle$, where the t_i are the basic primitive translations. By (12.17) and (13.24), the irreducible representations are all one-dimensional, and there are N_1N_2 or $N_1N_2N_3$ of them (according as the dimension of the space is 2 or 3). Based on the relations (18.7), the irreducible representations of T are those found in the proof of (12.17) with $n = 2$ or 3, ζ_j a primitive N_j^{th} root of 1, and $0 \le b_j < N_j$. Specifically, *one* such irreducible representation is

$$U\left(\sum_{j=1}^{n} a_j t_j\right) = \prod_{j=1}^{n} (\zeta_j^{b_j})^{a_j} \qquad \text{for } 0 \le a_j < N_j. \tag{18.8}$$

The general form of a representation U of T is sometimes given in the chemical literature (for each $a_j = 1$) as

$$\exp\left[i\frac{2\pi b_j}{t_j N_j}\cdot t_j\right] \qquad \text{for } 0 \le b_j \le N_j - 1,$$

where $\exp(A)$ denotes e^A and the dot denotes the scalar (dot) product of vectors.

Now the reducible representations of T may be formed from (18.8) in the usual way by taking linear combinations with integer coefficients of the irreducible representations.

We turn next to some specific space groups for lattices in $\mathbf{R}^2$. The five possible lattice units are shown in figure 11. We shall consider two of the seven groups associated with a rectangular lattice; the others are treated in standard references such as *International Tables for X-Ray Crystallography*, vol. 1. In figure 12 are shown the configurations that must be preserved by point operations and translations; each ⊢ or ⌊ must be carried onto another ⊢ or ⌊. The symbols *pm* and *pg* are often used to denote

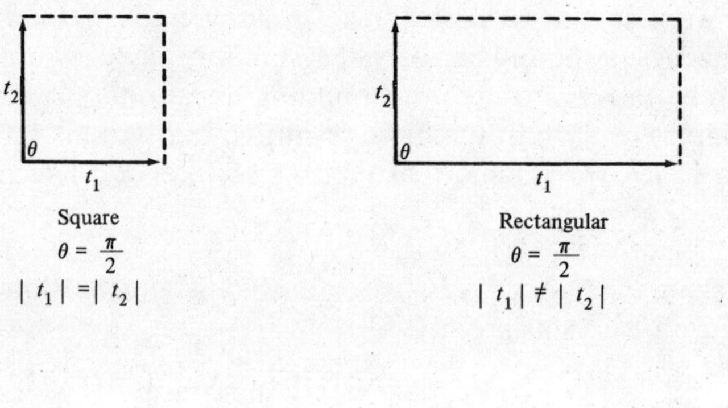

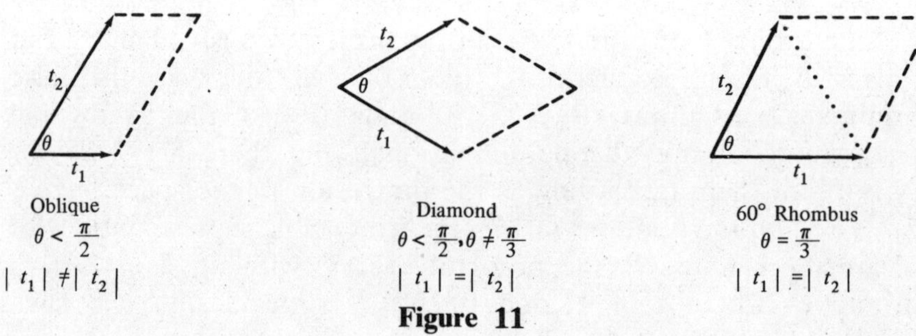

Figure 11

the groups of these two configurations (here p stands for "primitive," m for "mirror plane," and g for "glide plane"). Each of the groups pm and pg has a point group that is cyclic of order 2.

The group pm has a reflection σ_v generating its point group; it also has the basic primitive translations t_1 and t_2. Thus we may regard pm as generated by the elements $(\sigma_v, 0)$, (E, t_1), and (E, t_2). By inspection, one can see that $(\sigma_v, 0)$ commutes with (E, t_1) and that

$$(\sigma_v, 0)(E, t_2) = (E, -t_2)(\sigma_v, 0).$$

Figure 12

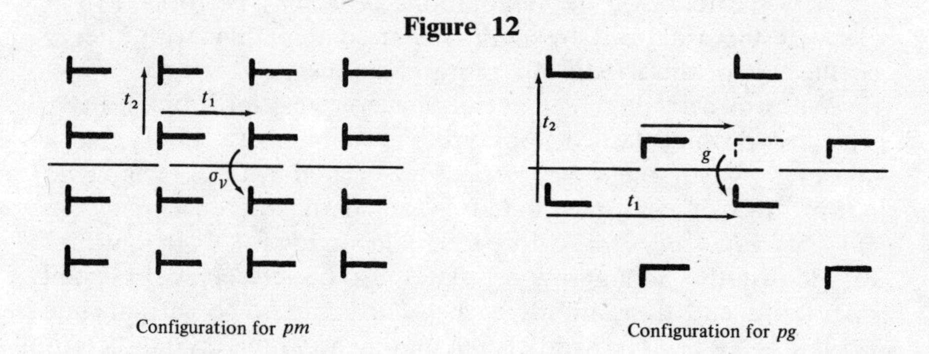

Configuration for *pm*   Configuration for *pg*

In order to write *pm* as an *abstract* group, we must change the primitive translation group T to multiplicative notation, that is, we write t_1t_2 instead of t_1+t_2, and t_1^{-1} in place of $-t_1$. Then we may summarize our discussion of the structure of *pm* as

$$pm=\langle s, t, u : s^2=1,\ st=ts,\ su=u^{-1}s,\ tu=ut\rangle,$$

where we have written s for σ_v, t for t_1, and u for t_2. Of course, *pm* as characterized above is infinite; if we adopt cyclic boundary conditions on t and u, we have

$$pm=\langle s, t, u : s^2=t^n=u^m=1,\ st=ts,$$
$$su=u^{-1}s,\ tu=ut\rangle,$$

for some (presumably large) integers n and m. Either way, $T=\langle t, u\rangle$ forms a normal subgroup, as shown before, since conjugation of the generators t and u of T by any generator of *pm* gives an element of T, by the defining relations. We note, however, that $\langle s\rangle$ is not normal since

$$u^{-1}su=su^2\notin\langle s\rangle.$$

The group *pg* has translations t_1 and t_2, and also a *glide plane* $(\sigma_v, \frac{1}{2}t_1)$ as shown in figure 12. Now by inspection of the geometry of the figure, we see that

$$(\sigma_v, \tfrac{1}{2}t_1)^2=(E, t_1) \tag{18.9}$$

and that

$$(\sigma_v, \tfrac{1}{2}t_1)(E, t_2)=(E, -t_2)(\sigma_v, \tfrac{1}{2}t_1), \tag{18.10}$$

which may be described in words as

> the product of two glide plane operations is a basic primitive translation in the horizontal direction

and

> a glide plane operation followed by a basic primitive translation *up* is the same as a primitive translation *down* followed by a glide plane operation.

However, if we attempt to verify (18.9) and (18.10) by means of the product rule (18.6), we have terms such as $\sigma_v\frac{1}{2}t_1$ whose meaning is not clear; the reason is that the glide plane is composed of two inseparable parts, the reflection and the half-translation, and moreover the latter is not primitive. The difficulty may be easily resolved if we resort to an abstract group presentation, again changing the primitive translation subgroup into multiplicative notation. Let g denote the glide plane operation and u the basic primitive translation upward; then (18.10) gives the relation $ug = gu^{-1}$, and (18.9) says that g^2 is the basic primitive translation to the right, so we need only two generators. Thus we may write

$$pg = \langle g, u : ug = gu^{-1} \rangle,$$

or if we assume cyclic boundary conditions, there are integers n and m such that

$$pg = \langle g, u : g^{2n} = u^m = 1, \; ug = gu^{-1} \rangle.$$

It is important to note that here the subgroup T of primitive translations is generated by g^2 and u. Now the point group pg/T is cyclic of order 2, and $pg = T \cup Tg$, but g does *not* generate a subgroup of order 2 in pg.

Thus we conclude that pm is symmorphic and pg is nonsymmorphic. Once we have introduced the idea of semi-direct products, we shall be able to say even a bit more about pg.

As a prelude to the final topic of this section, recall from the end of section 5 the *extension problem* for groups, which is: given groups N and H, what groups G exist for which $N \trianglelefteq G$ and $G/N \cong H$? (Strictly speaking, we ask that G contain an isomorphic copy of N, but such niceties of terminology can be dispensed with here.) The extension problem has a trivial answer for every N and H, namely $G = N \times H$, the external direct product from (12.5). Now we shall consider a more general product, in which the subgroup H of a product of-H-by-N is not necessarily a normal subgroup of the product. We make the idea precise in the following:

(18.11) DEFINITION. If $A \trianglelefteq G$, $H \leq G$, $G = HA$, and $H \cap A = \{1\}$, then G is called the *semi-direct product* of H and A.

If either H or A is the subgroup $\{1\}$, then the decomposition of G as a semi-direct product is trivial; we shall assume in general that H and A are both different from $\{1\}$.

Directly from theorem (5.10), we have:

(18.12) PROPOSITION. If G is the semi-direct product of subgroups H and A (with $A \trianglelefteq G$), then $H \cong G/A$.

Two examples will help to illustrate our definition. First, let

$$G = D_4 = \langle r, c : r^4 = c^2 = 1, \; cr = r^{-1}c \rangle,$$

and let $A = \langle r \rangle$ and $H = \langle c \rangle$. Then clearly G is the semi-direct product of H and A. On the other hand, we note that if we take $A = \langle r^2 \rangle$, the only *normal* subgroup of order 2, then there is *no* subgroup H for which we can write G as the semi-direct product of H and A since A is contained in every subgroup of order 4 (see figure 5 in section 5).

For our other example, consider

$$Q_2 = \langle x, y : x^4 = 1,\ x^2 = y^2,\ yx = x^{-1}y \rangle.$$

Now Q_2 has a unique subgroup $\langle x^2 \rangle$ of order 2, and it is contained in each of the subgroups $\langle x \rangle$, $\langle y \rangle$, and $\langle xy \rangle$ of order 4; hence Q_2 cannot be written as the direct product of a subgroup of order 4 with a normal subgroup of order 2. In addition, because Q_2 has only the subgroups listed above (together with itself and $\{1\}$), it cannot be written as the semi-direct product of a subgroup of order 2 with a normal subgroup of order 4.

An immediate consequence of definition (18.11) is the observation that every direct product is a semi-direct product. The example of D_4 above shows that a semi-direct product need not be a direct product.

Recall from exercise 7.10 that the automorphisms of a group form a group under the operation of composition. If G is the semi-direct product of H and A, we shall need to consider the group $\mathfrak{A}(A)$ of automorphisms of A in the following basic result on semi-direct products:

(18.13) PROPOSITION. Let G be the semi-direct product of H and A (with $A \trianglelefteq G$). Then:

(a) each $g \in G$ can be written uniquely as a product ha with $h \in H$ and $a \in A$;

(b) if $h_1, h_2 \in H$ and $a_1, a_2 \in A$, then

$$(h_1 a_1)(h_2 a_2) = (h_1 h_2)[(h_2^{-1} a_1 h_2) a_2];$$

(c) the function $\phi : H \to \mathfrak{A}(A)$ given by $\phi(h) : a \to hah^{-1}$ is a homomorphism;

(d) $\phi(h)$ in part (c) need not be an inner automorphism of A.

PROOF of (a). Since $G = HA$, we know that each element of G can be written as ha, with $h \in H$ and $a \in A$. If $ha = h'a'$, then

$$(h')^{-1} h = a' a^{-1} \in H \cap A,$$

so $(h')^{-1} h = 1$, whence $h' = h$; similarly, $a' = a$.

PROOF of (b). Since $A \trianglelefteq G$, we have

$$(h_1a_1)(h_2a_2) = h_1(h_2h_2^{-1})a_1h_2a_2$$
$$= (h_1h_2)[(h_2^{-1}a_1h_2)a_2] \in HA.$$

PROOF of (c). If h, $h' \in H$, then

$$\phi(hh'): a \to (hh')a(hh')^{-1} = h(h'ah'^{-1})h^{-1}$$

and

$$\phi(h)\phi(h'): a \to hh'ah'^{-1}h^{-1}.$$

PROOF of (d). In the example above with $G = D_4$ and $A = \langle r \rangle$, A is abelian, so its only inner automorphism is the identity, but $\phi(c): r \to crc^{-1} = crc = r^{-1}$, which is not the identity.

Recall from (12.6) that if G is the *direct* product of H and A, then every element in G can be written uniquely as a product ha with $h \in H$ and $a \in A$, and that the operation in G may be expressed as

(18.14) $$(h_1a_1)(h_2a_2) = (h_1h_2)(a_1a_2).$$

We may reconcile this product with the one in (18.13b) by observing that from (12.3) we know that h_2 commutes with a_1 when the product is direct; thus (18.13b) reduces to (18.14) for a direct product.

Given groups H and A, we can form an "external" semi-direct product by analogy with (12.5). By changing notation, if necessary, we take $H \cap A = \{1\}$. First we let ϕ be a homomorphism from H into $\mathfrak{A}(A)$ (noting that *each* $\phi(h)$ is then an automorphism of A) and set

(18.15) $$a^h = \phi(h^{-1})a \qquad \text{for } a \in A,\ h \in H.$$

Then

$$(a^h)^k = (\phi(h^{-1})a)^k = \phi(k^{-1})\phi(h^{-1})a$$
$$= \phi((hk)^{-1})a = a^{hk} \qquad \text{for } h,\ k \in H;$$
$$a^1 = \phi(1)a = a;$$

hence by (2.4) it follows that (18.15) defines an action of H on A. Now we take G to be the set of formal products of the form ha with $h \in H$ and $a \in A$. (No operation other than juxtaposition is assumed between h and a.) For h, $k \in H$ and a, $b \in A$, we define

(18.16) $$(ha)(kb) = (hk)(a^kb) \in HA = G.$$

Clearly $(ha)(1 \cdot 1) = h(a^1) = ha = (1 \cdot 1)(ha)$, so G has the identity element $1 \cdot 1$, which we shall write simply as 1. Since $\phi(h^{-1})$ is an automorphism of A, there is a unique solution b to $a^{h^{-1}}b = 1$ in A, and for this b, $h^{-1}b$ is an inverse for ha. For associativity, we observe that if $h, k, m \in H$ and $a, b, c \in A$,

$$\begin{aligned}[(ha)(kb)](mc) &= [(hk)(a^k b)](mc) = (hkm)(a^k b)^m c \\ &= (hkm)(a^{km} b^m c) = (ha)[(km)(b^m c)] \\ &= (ha)[(kb)(mc)],\end{aligned}$$

as required. Hence G is a group with the operation (18.16). If we identify A with the set of elements $1a$ (abbreviated a) and H with the set of elements $h1$ (abbreviated h), then A and H are clearly subgroups of G. Finally, for $h \in H$, $a \in A$,

$$\begin{aligned}h^{-1}ah &= (h^{-1}1)(1a)(h1) = [h^{-1}(1^1 a)](h1) \\ &= (h^{-1}a)(h1) = 1a^h \in A,\end{aligned}$$

so $A \trianglelefteq G$. This completes the proof that G is the semi-direct product of H and A. This external product may be related to the internal one by comparing the ϕ here with the ϕ of (18.13c).

Note that if in (18.15) $\phi(h^{-1})$ is the identity function for every $h \in H$, then the result in (18.16) is the direct product. In some instances, the *only* homomorphism from H to $\mathfrak{A}(A)$ is the trivial one carrying every $h \in H$ to the identity function; in this event the only semi-direct product of H and A is the direct product. For example, let $H = A \cong Z_3$; then since $\mathfrak{A}(Z_3) \cong Z_2$ and since when ϕ is a homomorphism, the order of $\phi(x)$ must divide the order of x, the only possible ϕ for (18.15) is the trivial one. Of course the only semi-direct product of H and A must be $Z_3 \times Z_3$ (see (12.15)).

To conclude the discussion, we observe first that *pm*, in either its finite form or its infinite form, is the semi-direct product of $\langle s\rangle$ with $T = \langle t, u\rangle$. The reader can easily reconcile the different notations (18.6) and (18.16) for the group product. Second, we show:

(18.17) PROPOSITION. The space group *pg* cannot be written as a semi-direct product of some subgroup with $T = \langle g^2, u\rangle$.

PROOF. We take $pg = \langle g, u : ug = gu^{-1}\rangle$ since even with the cyclic boundary condition we would assume that the orders of g and u were very large (for a lattice of atoms in a crystal). We know that *pg* is not the semi-direct product of $\langle g\rangle$ and T since $g^2 \in T$; the point is to show that there is *no* subgroup H of *pg* such that $pg = HT$, $H \cap T = 1$. But $[pg : T] = 2$, so such an H would have to have order 2, and we know that no element of pg other than the identity has finite (or at least small finite) order. Therefore, pg cannot be written as a semi-direct product with T as the normal subgroup.

However, as our final observation, let us establish the following:

(18.18) PROPOSITION. The space group pg can be written as a semidirect product with $\langle u \rangle$ as the normal subgroup.

PROOF. First we show that $\langle u \rangle$ is a normal subgroup of pg. Now

$$g^{-1}ug = g^{-1}(gu^{-1}) = u^{-1}, \tag{18.19}$$

and (as one can always do from an equation like the preceding)

$$g^{-1}u^{-1}g = (g^{-1}ug)^{-1} = u; \tag{18.20}$$

hence $\langle u \rangle \trianglelefteq pg$. Now since $gu \neq ug$, pg cannot be the *direct* product of $\langle g \rangle$ and $\langle u \rangle$, but if we can show that $pg = \langle g \rangle \langle u \rangle$, it will follow (with $\langle g \rangle \cap \langle u \rangle = \{1\}$) that pg is the semi-direct product of $\langle g \rangle$ and $\langle u \rangle$. Note that $pg = \langle g, u \rangle$ does *not* imply immediately that $pg = \langle g \rangle \langle u \rangle$. Now since we have the defining relation $ug = gu^{-1}$, we see that any u in a product of powers of g and powers of u can be moved to the right across any g^k with $k > 0$. From (18.20) we have $u^{-1}g = gu$, so any u^{-1} in such a product may also be moved to the right across any g^k with $k > 0$. (Of course u changes to u^{-1} in such a move, and u^{-1} changes to u, but in either event the $u^{\pm 1}$ moves to the right.) Now from (18.20) we have $ug^{-1} = g^{-1}u^{-1}$, and from (18.19) we have $u^{-1}g^{-1} = g^{-1}u$; hence any occurrence of $u^{\pm 1}$ may be moved to the right in a product across any g^k with $k < 0$. Hence in any arbitrary product of powers of g and u, all powers of u may be collected on the right, leaving all powers of g on the left, that is, the element may be brought into the form $g^j u^k$ with integer exponents. Therefore, $pg = \langle g \rangle \langle u \rangle$. This completes the proof.

Note that *a space group is symmorphic if and only if it can be written as the semi-direct product of its point group and its subgroup of primitive translations.* Thus pg remains nonsymmorphic despite (18.18).

EXERCISES

18.1. The square bipyramid also has a *full symmetry group* G^* consisting of the group G found in section 1 (note that $G \cong D_4$) and a coset

Gz whose representative z is an inversion of all vertices through the center of the solid. Find the character table of G^*.

18.2. From the character tables found in section 14, observe that (18.4) cannot hold with $H = A_4$, $G = S_4$. Determine why (18.4) does not apply, and discuss this result geometrically with respect to the fact that (by exercise 7.6) A_4 is the group of rigid symmetries of the tetrahedron.

18.3. In the notation of the text, verify geometrically that

$$C_4 t_2 = -t_1,$$
$$C_4 t_1 + t_2 = 2t_2,$$
$$(C_4, t_2)(C_4, t_1) = (C_4^2, 2t_2),$$
$$(C_4, t_1 - t_2)^{-1} = (C_4^3, -C_4^3(t_1 - t_2)).$$

18.4. We found in (18.18) that pg can be written as a semi-direct product of $\langle g \rangle$ and $\langle u \rangle$. Determine whether, for *its* vertical basic primitive translation u, the group pm can be written as a semi-direct product of some subgroup and $\langle u \rangle$. Can it be written as a *direct* product of some subgroup and $\langle u \rangle$?

18.5. We know that $S_4 = \langle r, x \rangle$ (as in the text following (18.4)) but that $S_4 \neq \langle r \rangle \langle x \rangle$. Examine the proof of (18.18) and determine why one cannot carry out for S_4 the argument given there to show that $pg = \langle g \rangle \langle u \rangle$.

18.6. Determine the two distinct semi-direct products of Z_2 by Z_∞.

Section 19
Some Infinite Groups Used in Physics

We have already encountered various infinite groups, notably Z_∞, $GL(n, \mathbf{C})$, and $SL(n, \mathbf{C})$. We shall have occasion now to think of the infinite cyclic group in physical (geometrical) terms as a rotation through angle $2\pi/b$, where b is an irrational number. This section presents some infinite groups that are of special interest for applications in physics and, in particular, examines the relationship between the unitary unimodular group of degree 2 and the group of rotations in $\mathbf{R}^3$.

An immediate generalization of what we have done thus far is the concept of *continuous groups of transformations.* The most familiar examples of such transformations are rotations. Recall that a finite cyclic group Z_n may be thought of as the rotations in $\mathbf{R}^2$ of a regular n-sided polygon; a generator of Z_n is a rotation through $2\pi/n$. We may write $Z_n = \langle T_1 \rangle$ and let

$$(19.1) \qquad T_k = T_1^k \quad \text{for} \quad k = 0, 1, \ldots, n-1,$$

noting that $T_1^n = T_1^0$ is the identity. The situation described by continuous groups is that in which n approaches infinity, and the polygon becomes a circle. The index k in (19.1) is no longer applicable since between any two rotations there is another rotation; the index now varies continuously over $\mathbf{R}$ and is replaced by a continuously varying parameter. We may assign the parameter at will, and the usual choice for rotations is the obvious one of the angle of rotation. We can write the transformations

$$x' = x \cos\theta + y \sin\theta$$

$$y' = -x \sin\theta + y \cos\theta$$

of $\mathbf{R}^2$ in terms of the matrix

$$(19.2) \qquad R_2(\theta) = \begin{bmatrix} \cos\theta & \sin\theta \\ -\sin\theta & \cos\theta \end{bmatrix},$$

as in fact we did at (9.2) for $\theta = \pi/2$. Clearly,

$$(19.3) \qquad R_2(0) = R_2(2\pi) = I_2,$$

and $R_2(\theta_1)R_2(\theta_2)=R_2(\theta_1+\theta_2)$. The transformations defined by (19.2) form the *rotation group* $O^+(2)$ in $\mathbf{R}^2$.

In terms of linear algebra, a matrix is *orthogonal* if its inverse is equal to its transpose. If A is an orthogonal matrix, then $|A|=\pm 1$ since from the theory of determinants, $|A^t|=|A|$. Now if $A\in GL(2, \mathbf{R})$ is orthogonal and $|A|=1$, then A can be written in the form (19.2) for some θ, and if $|A|=-1$, then A can be written as

$$\begin{bmatrix} 1 & 0 \\ 0 & -1 \end{bmatrix} R_2(\theta)$$

for some θ (see exercise 19.1). The orthogonal matrices in $GL(n, \mathbf{R})$ form the *orthogonal group* $O(n)$, and the orthogonal matrices with determinant 1 form the normal subgroup $O^+(n)$; if $n>2$, $O^+(n)$ is called the *rotation group* and $O(n)$ the *rotation-inversion group* in $\mathbf{R}^n$. For further discussion, see Curtis [**3**], p. 270.

In (19.1) we had the finite group Z_n generated by the rotation T_1; we would like to preserve the idea of a generator in dealing with a continuous group. Let θ in (19.2) be an infinitesimal quantity ε. Then we may say

$$R_2(\varepsilon)=\begin{bmatrix} 1 & \varepsilon \\ -\varepsilon & 1 \end{bmatrix}=I_2+\varepsilon\begin{bmatrix} 0 & 1 \\ -1 & 0 \end{bmatrix}, \tag{19.4}$$

and in general, for any continuous group, we can write the generators T_i in the neighborhood of the group identity in the form

$$T_i=1+\varepsilon_i t_i. \tag{19.5}$$

The finite transformations are then generated by iterating the infinitesimal ones. Suppose for example that we wish to reconstruct (19.2) from a knowledge of (19.4). Divide the finite angle θ into N small steps, with N "eventually" infinite: thus $\varepsilon=\theta/N$. Then we can define $R_2(\theta)$ by an iteration

$$R_2(\theta)=\lim_{N\to\infty}[R_2(\theta/N)]^N,$$

which by (19.4) becomes

$$R_2(\theta)=\lim_{N\to\infty}\left[I_2+\frac{\theta}{N}r_2\right]^N=e^{\theta r_2}; \tag{19.6}$$

the identity of the limit and the power series for the exponential can be established formally by the same proof as that used in elementary calculus, even when the quantity r_2 is a matrix (see exercise 19.3).

Just as the study of cyclic groups alone does not provide much insight into the structure of finite groups in general, continuous groups generated by a single element (called *one-parameter groups*) do not exhaust the subject of continuous groups. In general we need r parameters, and the group is generated by the infinitesimal transformations

$$T_i(\varepsilon_i) = 1 + \varepsilon_i t_i \quad \text{for} \quad i = 1, 2, \ldots, r. \tag{19.7}$$

These transformations in the immediate neighborhood of the group identity 1 generate the *infinitesimal subgroup* of the continuous group. A continuous group is not in general abelian, but for distinct i and j,

$$T_i(\varepsilon_i)T_j(\varepsilon_j) = 1 + \varepsilon_i t_i + \varepsilon_j t_j [+\varepsilon_i \varepsilon_j t_i t_j],$$

where we may suppress the "very small" term in brackets and, in doing so, claim that (for $i \neq j$)

$$T_i(\varepsilon_i)T_j(\varepsilon_j) = T_j(\varepsilon_j)T_i(\varepsilon_i).$$

In this sense, one says that the infinitesimal subgroup is abelian, though we shall qualify this statement subsequently.

What we have done for the infinitesimal neighborhood of the identity transformation 1 can be repeated in the infinitesimal neighborhood of *any* finite transformation such as $R_2(\theta)$. This observation suggests that the way to explore the structure of a continuous group is to study the infinitesimal neighborhood of an arbitrary finite transformation and thereby to cover the entire group with copies of the infinitesimal subgroup, as was done above in the trivial example of $O^+(2)$. That this approach is possible was shown by Sophus Lie (1842–1899), whose work dominates the theory of continuous groups to this day. We shall not develop the theory here, but our discussion will in fact draw upon ideas of Lie groups in a naive setting.

To consider more general elements of the infinitesimal subgroup, we take a product of r transformations of the form (19.7), with distinct i and with each $\varepsilon_i \geq 0$; we obtain (under the claim that distinct generators commute) the form

$$1 + \sum_{i=1}^{r} \varepsilon_i t_i \tag{19.8}$$

for this product. We need also to consider the powers of a single generator $T_i(\varepsilon_i)$; for $n > 1$,

$$(T_i(\varepsilon_i))^n = 1 + n\varepsilon_i t_i + \frac{n(n-1)}{2}\varepsilon_i^2 t_i^2 + \dots.$$

If n is large, the third term may not be negligible, but for large n, we regard $(T_i(\varepsilon_i))^n$ as a finite transformation, that is, as one outside the infinitesimal subgroup; hence we shall take the relevant powers of a single $T_i(\varepsilon_i)$ to be $T_i(n\varepsilon_i)$. Thus (19.8) gives the general form for *all* elements of the infinitesimal subgroup.

Now in moving away from the identity transformation 1 by means of the product $T_k(\varepsilon_k)T_j(\varepsilon_j)$, we suppressed the higher order term in $t_k t_j$, but in fact the successive transformations $T_j(\varepsilon_j)$, $T_k(\varepsilon_k)$, $T_j(-\varepsilon_j)$, $T_k(-\varepsilon_k)$ will not quite bring us back to the identity, as commutativity would have it. The displacement from 1 effected by the four consecutive transformations

$$T_k(-\varepsilon_k)T_j(-\varepsilon_j)T_k(\varepsilon_k)T_j(\varepsilon_j)$$

brings one back close to the identity and therefore may be thought of as an element of the form (19.8). From (19.6) we now take the first *three* terms of the Taylor series for e^z—rather than the linear Taylor polynomial (19.7)—and write

$$T_j(\varepsilon_j) = 1 + \varepsilon_j t_j + \tfrac{1}{2}\varepsilon_j^2 t_j^2.$$

A calculation suppressing terms of higher order than the second (left as exercise 19.4) leads to

$$\text{(19.9)} \qquad T_k(-\varepsilon_k)T_j(-\varepsilon_j)T_k(\varepsilon_k)T_j(\varepsilon_j) = 1 - \varepsilon_j\varepsilon_k(t_j t_k - t_k t_j).$$

(Note the similarity in form between the left-hand side of (19.9) and the commutator $ghg^{-1}h^{-1}$ introduced in (12.19).) In order to write the product of (19.9) in the form (19.8), we require constants $c_{jk}^{(i)}$ such that

$$\text{(19.10)} \qquad t_j t_k - t_k t_j = \sum_{i=1}^{r} c_{jk}^{(i)} t_i \quad \text{for} \quad 1 \le j,\, k \le r.$$

The left-hand side of (19.10) is called the *Lie product* of t_j and t_k.

Now just as we were able in section 7 to give a full specification of a group by means of a presentation in terms of generators and relations, so the set of generators (19.7) together with the equations (19.10) completely defines the structure of a continuous group. The constants $c_{jk}^{(i)}$ are called the *structure constants* of the group.

Now consider the rotation group $O^+(3)$ of $\mathbf{R}^3$. We may regard the rotation (19.2) as taking place in $\mathbf{R}^3$ if we think of it as a rotation about the Z-axis instead of a rotation about the origin in $\mathbf{R}^2$. Then each point on the Z-axis is carried to itself, and the matrix of the linear transformation is

$$R_z(\theta)=\begin{bmatrix} \cos\theta & \sin\theta & 0 \\ -\sin\theta & \cos\theta & 0 \\ 0 & 0 & 1 \end{bmatrix}.$$

(Note that we have taken a special case of exercise 19.2.) The corresponding infinitesimal transformation is represented by

$$R_z(\varepsilon_z)=1+\varepsilon_z r_z \quad \text{with} \quad r_z=\begin{bmatrix} 0 & 1 & 0 \\ -1 & 0 & 0 \\ 0 & 0 & 0 \end{bmatrix},$$

where we have written 1 for the identity I_3.

Similarly, we take $R_x(\varepsilon_x)=1+\varepsilon_x r_x$ and $R_y(\varepsilon_y)=1+\varepsilon_y r_y$, where

$$r_x=\begin{bmatrix} 0 & 0 & 0 \\ 0 & 0 & 1 \\ 0 & -1 & 0 \end{bmatrix} \quad \text{and} \quad r_y=\begin{bmatrix} 0 & 0 & -1 \\ 0 & 0 & 0 \\ 1 & 0 & 0 \end{bmatrix}.$$

Now we can use this matrix representation to derive the structure constants of the group. Using the usual notation $[a, b]=ab-ba$ of the Lie product, we find

$$(19.11) \qquad [r_x, r_y]=-r_z, \qquad [r_y, r_z]=-r_x, \qquad [r_z, r_x]=-r_y;$$

hence in this case, all of the structure constants are either 0 or -1. The relations (19.11) characterize the group and not merely the particular set of matrices we have used to represent it; *all* matrix representations of the group obey the same relations.

The commutation relations (19.11) give us some insight into the geometrical meaning of the general relation (19.9); for example, if the transformations are infinitesimal rotations around the X- and Y-axes, then

$$(19.12) \qquad \begin{cases} R_y(-\varepsilon_y)R_x(-\varepsilon_x)R_y(\varepsilon_y)R_x(\varepsilon_x)=1-\varepsilon_x\varepsilon_y[r_x, r_y] \\ \qquad\qquad =1+\varepsilon_x\varepsilon_y r_z \\ \qquad\qquad =R_z(\varepsilon_x\varepsilon_y), \end{cases}$$

and the result of the four "small" rotations around the X- and Y-axes is a "very small" rotation around the Z-axis. This result may be qualitatively demonstrated by means of two pins, four hands, and a tennis ball.

Now we return to linear algebra to give a characterization of the unitary unimodular group of degree 2. If $A \in GL(n, \mathbf{C})$, we define the *conjugate transpose* A^* of A to be the transpose of the matrix whose entries are, respectively, the complex conjugates of the entries of A. Clearly, the operations of taking the transpose and taking thc conjugates may be done in either order. Moreover, since transposing twice returns the matrix to its original form, and similarly for conjugation, we have $(A^*)^* = A$.

If $A \in GL(n, \mathbf{C})$ and $A^* = A^{-1}$, then A is called *unitary*. If $A^* = A^{-1}$ and $B^* = B^{-1}$, then

$$(A^{-1})^{-1} = A = (A^*)^* = (A^{-1})^*$$

and

$$(AB)^{-1} = B^{-1}A^{-1} = B^*A^* = \bar{B}^t\bar{A}^t = (\overline{AB})^t = (AB)^*.$$

We conclude that the set $U(n)$ of unitary matrices of degree n forms a group under matrix multiplication.

Now the determinant of a matrix A is merely a sum of products of the entries, and hence by (13.16) and the fact that $|A^t| = |A|$, we see that $|A^*| = \overline{|A|}$. If $A \in U(n)$, then

$$1 = |I_n| = |A||A^{-1}| = |A||A^*| = |A|\overline{|A|},$$

and hence the complex norm of $|A|$ is 1. In particular, there is a real θ such that $|A| = e^{i\theta}$. From here one can easily prove the following characterization of $U(2)$:

(19.13) PROPOSITION. If $A \in U(2)$, then there exist a, $b \in \mathbf{C}$ and $\theta \in \mathbf{R}$ such that

$$A = \begin{bmatrix} 1 & 0 \\ 0 & e^{i\theta} \end{bmatrix} \begin{bmatrix} a & b \\ -\bar{b} & \bar{a} \end{bmatrix}$$

and $|a|^2 + |b|^2 = 1$.

Now we shall consider the subgroup $SU(2)$ of $U(2)$ consisting of the unitary matrices having determinant 1; $SU(2)$ is called the *unitary unimodular group* of degree 2 (the term has its origin in the fact that $SL(n, \mathbf{C})$ is sometimes called *unimodular* and that $SU(n) = SL(n, \mathbf{C}) \cap U(n)$). In this case, the matrix A of (19.13) reduces to

(19.14) $$A = \begin{bmatrix} a & b \\ -\bar{b} & \bar{a} \end{bmatrix} \quad \text{with} \quad |a|^2 + |b|^2 = 1.$$

Now since a and b are complex numbers, an analog of what we did in (19.2) for orthogonal matrices is to set

$$a = e^{i\xi} \cos \eta, \qquad b = e^{i\zeta} \sin \eta, \quad \text{for} \quad \xi, \eta, \zeta \in \mathbf{R}.$$

Since $\bar{a} = e^{-i\xi} \cos \eta$ and $\bar{b} = e^{-i\zeta} \sin \eta$, we have

$$|a|^2 + |b|^2 = a\bar{a} + b\bar{b} = \cos^2 \eta + \sin^2 \eta = 1,$$

as required, and hence

$$A = \begin{bmatrix} e^{i\xi} \cos \eta & e^{i\zeta} \sin \eta \\ -e^{-i\zeta} \sin \eta & e^{-i\xi} \cos \eta \end{bmatrix}. \tag{19.15}$$

This formula gives a general form for the matrices in $SU(2)$ in terms of three real parameters.

Now we shall consider the matrices of $SU(2)$ in terms of the characterization (19.14). The discussion preceding (19.12) encourages us to approach a matrix $A \in SU(2)$ by way of its infinitesimal transformations. We have seen that A is determined by three real parameters; in this case we write A in terms of ε_1, ε_2, ε_3 as

$$S(\varepsilon_1, \varepsilon_2, \varepsilon_3) = \begin{bmatrix} 1 + \frac{1}{2} i \varepsilon_3 & \frac{1}{2} i (\varepsilon_1 - i\varepsilon_2) \\ \frac{1}{2} i (\varepsilon_1 + i\varepsilon_2) & 1 - \frac{1}{2} i \varepsilon_3 \end{bmatrix}. \tag{19.16}$$

Clearly $S(\varepsilon_1, \varepsilon_2, \varepsilon_3)$ satisfies (19.14) to order ε. We may rewrite (19.16) as

$$(19.17) \begin{cases} S(\varepsilon_1, \varepsilon_2, \varepsilon_3) = 1 + \sum_{j=1}^{3} \varepsilon_j s_j, \quad \text{where} \\ s_1 = \frac{1}{2} \begin{bmatrix} 0 & i \\ i & 0 \end{bmatrix}, \quad s_2 = \frac{1}{2} \begin{bmatrix} 0 & 1 \\ -1 & 0 \end{bmatrix}, \quad s_3 = \frac{1}{2} \begin{bmatrix} i & 0 \\ 0 & -i \end{bmatrix}. \end{cases}$$

The reason for the apparently arbitrary factors of $\frac{1}{2}$ may be seen if we write down the structure relations of $SU(2)$. We find that from (19.16) and (19.17) as given,

$$[s_1, s_2] = -s_3, \qquad [s_2, s_3] = -s_1, \qquad [s_3, s_1] = -s_2, \tag{19.18}$$

which are the same as the structure relations for $O^+(3)$, and thus the two infinitesimal groups have exactly the same structure. We have thus found a representation afforded by $\mathbf{C}^2$ for the infinitesimal part of $O^+(3)$.

The exploration of all the infinitesimal neighborhoods, however, does not tell us all about a space, that is, *local* equivalence may not insure *global* equivalence. A simple example is the comparison of the surfaces of the Möbius strip and a simple band (see figure 13), which would appear the same if one looked only at small neighborhoods of points on the surfaces but which are clearly distinct in the whole. Our last task is to see whether the *local* identity of $SU(2)$ and $O^{+}(3)$ extends to a *global* identity. In so doing, we shall see the meaning of the factors of $\frac{1}{2}$ that had to be introduced into (19.16) in order to make the structure relations in (19.18) come out to resemble (19.11).

We set $\varepsilon_1 = \varepsilon_2 = 0$ and employ the approach that led to (19.6) in order to find the finite transformation in $SU(2)$ that corresponds to a rotation through an angle θ about the Z-axis in $O^{+}(3)$. For the finite transformation we have

$$S_3(\theta) = \lim_{N\to\infty}\left[1 + \frac{\theta}{N} s_3\right]^N$$

$$= \lim_{N\to\infty}\begin{bmatrix} 1 + \dfrac{i\theta}{2N} & 0 \\ 0 & 1 - \dfrac{i\theta}{2N} \end{bmatrix}^N$$

$$= \lim_{N\to\infty}\begin{bmatrix} (1 + i\theta/2N)^N & 0 \\ 0 & (1 - i\theta/2N)^N \end{bmatrix},$$

whence, as in (19.6),

$$S_3(\theta) = \begin{bmatrix} e^{i\theta/2} & 0 \\ 0 & e^{-i\theta/2} \end{bmatrix}.$$

Now let $\theta = 2\pi$; then $S_3(2\pi) = -1$, and *it is not until* $\theta = 4\pi$ that S_3 is once more equal to the identity. The analogy to the Möbius strip is here quite clear: after a circuit of 2π, one is on the opposite "side" from the starting point, and the starting point is not reached

Figure 13

Möbius Strip

Simple Band

until after a circuit of 4π. Thus the matrices $S_3(\theta)$ form a double-valued representation of $O^+(3)$ while, of course, forming a single-valued one of $SU(2)$. (That is, the kernel of the homomorphism from $SU(2)$ to $O^+(3)$ has order 2.) In this situation $SU(2)$ is called the *covering group* of $O^+(3)$. The connection between these two groups is explored in greater detail in exercise 19.9.

We see from the foregoing discussion that there is a mathematical sense in which a rotation through 4π is "more like" the identity transformation than is a rotation through 2π. The reader who doubts the possibility of a realization in the real world may construct the toy illustrated in figure 14. If the ends of the three strings attached to the cardboard triangle are held fast and the triangle is rotated through 4π about the axis A, a little patience will enable one to untangle the strings while holding the triangle stationary. After a rotation of only 2π, however, it is impossible to untangle the strings. That this interesting mechanical fact is actually an illustration of the mathematics discussed above follows from a detailed topological discussion (see Bolker [**1**]).

In quantum mechanics the two-dimensional complex vectors that transform under $SU(2)$ are used to specify the spin states of certain elementary particles. The vector does not return to its original value under such a rotation. Most spin quantities are expressed as bilinear forms in the spin vectors and are therefore unchanged under the rotation. But if we imagine a physicist (or even a mathematician) living in a pivoted room with windows to

Figure 14

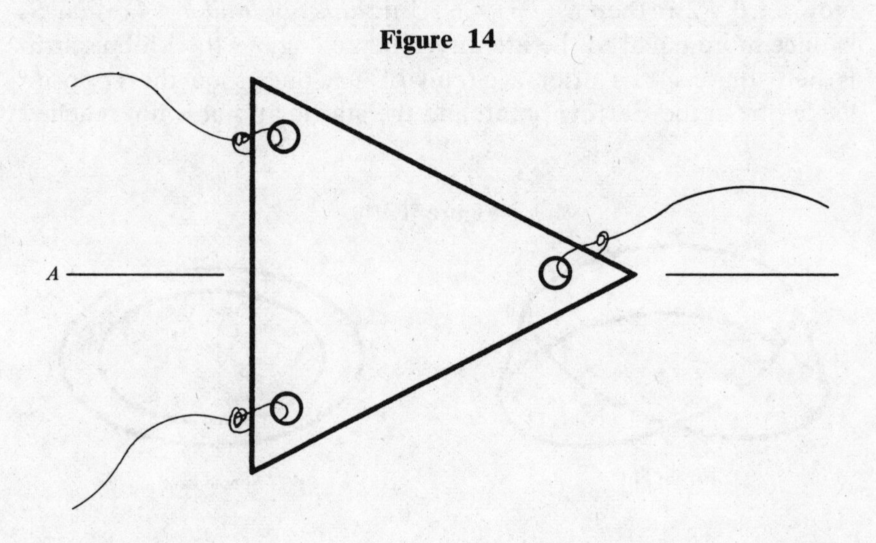

the outside world it is still an open question whether there is any experiment he could possibly perform in order to determine whether the room had been slowly rotated through an *odd* number of complete revolutions as he slept.* An *even* number of rotations is believed to be undetectable.

EXERCISES

19.1. Let $A \in O(2)$; prove if $|A| = 1$, then A can be written in the form (19.2), and if $|A| = -1$, then A can be written as the product of a reflection in the X-axis and a rotation of the form (19.2).

19.2. Let $A \in GL(3, \mathbf{R})$ and $|A| = 1$; prove that by a suitable choice of basis for $\mathbf{R}^3$, A can be written in the form

$$\begin{bmatrix} \cos\theta & \sin\theta & 0 \\ -\sin\theta & \cos\theta & 0 \\ 0 & 0 & 1 \end{bmatrix}.$$

19.3. Show by expanding the exponential that (19.6) is the same as (19.2).

19.4. Carry out the calculation for (19.9).

19.5. Verify the commutation relations in (19.11).

19.6. Explain why you would expect to find an infinite number of irreducible representations, having unbounded degrees, for every continuous group. (The statement is, in fact, true.)

19.7. Prove (19.13).

19.8. Prove that $U(2) \cap GL(2, \mathbf{R}) = O(2)$.

19.9. In order to see the connection of $SU(2)$ with $O^+(3)$ in more detail, we observe that they are afforded by $\mathbf{C}^2$ and $\mathbf{R}^3$, respectively. Let an element of $O^+(3)$ transform a real vector (x, y, z) into (x', y', z') with

$$x'^2 + y'^2 + z'^2 = x^2 + y^2 + z^2, \tag{19.19}$$

and let a matrix in $SU(2)$ transform a complex vector (ξ, η) into (ξ', η') with

$$|\xi'|^2 + |\eta'|^2 = |\xi|^2 + |\eta|^2. \tag{19.20}$$

Now represent (x, y, z) by

$$x = \xi\bar{\eta} + \bar{\xi}\eta, \qquad y = i(\xi\bar{\eta} - \bar{\xi}\eta), \qquad z = |\xi|^2 - |\eta|^2,$$

* A pendulum or gyroscope could of course be arranged so as to leave a trace of the *act* of turning, but back in their original orientations in the morning, they give no indication of having turned. It is this kind of evidence that is meant here.

and show that $x^2+y^2+z^2=(|\xi|^2+|\eta|^2)^2$ so that (19.20) implies (19.19). Finally, subject the vector (ξ, η) to the transformation $1+\varepsilon s_3$ and show that this induces a rotation of the vector (x, y, z) through an angle ε. If not convinced, try another infinitesimal transformation. (Since things that transform like a product of vectors are called tensors, one may say that vectors in $\mathbf{R}^3$ are tensors in the space of $SU(2)$. Vectors in that space, on account of their use in physics, are called *spinors.*)

19.10. The two-valued nature of the mapping of $O^+(3)$ into $SU(2)$ was shown only for the transformation $S_3(\theta)$. Show that it also holds for the other two transformations of $SU(2)$ generated by iterations of $S(\varepsilon_1, 0, 0)$ and $S(0, \varepsilon_2, 0)$.

19.11. A more artificial but purely algebraic illustration of the connection between $O^+(3)$ and $SU(2)$ may be obtained to show the global nature of the two-valued representation. One can prove that an arbitrary rotation in $O^+(3)$ may be expressed as a rotation through angle ψ about the Z-axis, followed by one through angle θ about the X-axis, and then followed by one through angle ϕ about the Z-axis. Thus any element of $O^+(3)$ may be decomposed as a product

$$\begin{bmatrix} \cos\psi & \sin\psi & 0 \\ -\sin\psi & \cos\psi & 0 \\ 0 & 0 & 1 \end{bmatrix} \begin{bmatrix} 1 & 0 & 0 \\ 0 & \cos\theta & \sin\theta \\ 0 & -\sin\theta & \cos\theta \end{bmatrix} \begin{bmatrix} \cos\phi & \sin\phi & 0 \\ -\sin\phi & \cos\phi & 0 \\ 0 & 0 & 1 \end{bmatrix},$$

which we denote by $A_{\psi,\theta,\phi}$. Now for a matrix $A \in SU(2)$ written as in (19.15), let

$$\theta = 2\eta, \qquad \phi = \xi + \zeta - \pi/2, \qquad \psi = \xi - \zeta + \pi/2,$$

and take $T(A) = A_{\psi,\theta,\phi}$. Show that T is a representation of $SU(2)$ and that the kernel of T has order 2.

Section 20

The Burnside Counting Theorem

A pure group-theoretic result of Burnside has been shown by G. Pólya and others to have applications to counting problems. This section illustrates the applications and also shows how Burnside's result ties in with characters as discussed in section 15. First, let us state the Counting Theorem and examine a proof that is essentially due to Pólya:

(20.1) THEOREM (Burnside [**2**], p. 191). Let G be a group acting on a point set Ω, and for each $g \in G$, let $\chi(g)$ be the number of points of Ω left fixed by g. Then

$$\frac{1}{|G|} \sum_{g \in G} \chi(g) \tag{20.2}$$

is equal to the number of distinct orbits of Ω under G.

PROOF. In the notation of (2.4), we count the number N of α^g ($\alpha \in \Omega$, $g \in G$) for which $\alpha = \alpha^g$. For fixed g, the number of such α^g is $\chi(g)$, and hence

$$N = \sum_{g \in G} \chi(g).$$

On the other hand, for fixed α, the number of such α^g is $|G_\alpha|$; thus if we write Ω as the union of distinct orbits α_j^G for $1 \leq j \leq r$, then

$$\begin{aligned} N = \sum_{\alpha \in \Omega} |G_\alpha| &= \sum_{j=1}^{r} \sum_{\alpha \in \alpha_j^G} |G_\alpha| \\ &= \sum_{j=1}^{r} |\alpha_j^G||G_{\alpha_j}| \qquad \text{by (5.4)} \\ &= \sum_{j=1}^{r} |G| = r|G| \qquad \text{by (3.11).} \end{aligned}$$

Therefore $r = \dfrac{1}{|G|} \sum_{g \in G} \chi(g)$, as was to be proved.

Theorem (20.1) can be simply illustrated with the dihedral group D_4. When viewed as the rigid symmetries of the square bipyramid, D_4 acts on a set of six points having two orbits. In the notation of section 1, the identity e fixes all six points; the rotations r, r^2, r^3 each fix only the two points 1 and 6; the motions rc and r^3c fix no points; and c and r^2c each fix two points. Thus (20.2) becomes

$$\tfrac{1}{8}(6+3(2)+2(0)+2(2))=2.$$

In contrast, when viewed as the symmetries of the square, D_4 acts on a set of four points having a single orbit. Using the same notation as above, e fixes four points, c and r^2c fix two points each, and the other five elements fix no points. Then (20.2) becomes

$$\tfrac{1}{8}(4+2(2)+5(0))=1.$$

For an application to counting problems let us consider an example that can be readily confirmed. The ammonia molecule NH_3 has the form of a trigonal pyramid with an equilateral base and isosceles sides (the upper half of the figure in exercise 1.5). Zero to three of the hydrogens at the vertices of the base may be replaced by chlorines; how many visually distinct molecules result? The rigid symmetries of this pyramid are precisely the rotations of the base; thus the group is cyclic of order 3. The point set Ω now consists of the $2^3=8$ *configurations* one can make by choosing H or Cl at each of the three vertices of the base. A rotation r through $2\pi/3$ will leave fixed only the two configurations having H at all three positions, or Cl at all three; similarly for r^2. The identity leaves fixed all eight configurations, that is, if *no movement* of the triangle is permitted then "H at the upper vertex and Cl at the lower two" is *different from* "H at the lower left vertex and Cl at the other two." Thus (20.2) becomes $\frac{1}{3}(8+2+2)=4$ orbits, which is to say that the eight configurations reduce to four visually distinct ones. This result is immediately verified when one realizes that the molecule (that is, the visually distinct configuration) is completely determined by the number of Cl substitutions: 0, 1, 2, or 3.

If, in the preceding example, we allow a substitution of a Cl or of a CH_3 at each H, a similar argument to the above gives $\frac{1}{3}(27+3+3)=11$ distinct configurations, a result that is a bit harder to verify in one's head. Note that

(20.3)
```
     H                  Cl
    / \                / \
  Cl———CH3   and    H———CH3
```

are distinct since the pyramid cannot be inverted; the patterns (20.3) correspond in a theoretical context to *optical isomers.* For experimental work, inversion of the pyramid (which is geometrically equivalent to turning a solid pyramid inside out) would enable us to transform one isomer into the other and would reduce the number of molecules to ten. If we allow inversion, the group of transformations has order 6 and is isomorphic to $D_3 = \langle r, c\rangle$. Each of the reflections will fix any configuration of the form

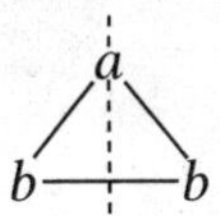

where we have three free choices for a and three for b. We still have $|\Omega| = 27$; thus (20.2) becomes $\frac{1}{6}(27+2(3)+3(9)) = 10$, in agreement with the preceding.

The discussion here is an illustration of how group theory can be used to tell what will *not* occur, a characterization expressed by some chemists. Specifically, the argument just concluded tells that from a mathematical point of view, *not more than* ten distinct molecules can result from substitution of Cl or CH_3 for H's in the ammonia molecule. Whether all ten such distinct molecules exist is a question for the chemist.

The benzene molecule C_6H_6 may be thought of as a hexagon in the plane with a C at each vertex and an H attached to each C and available for substitution. For example, one of the halogens F, Cl, Br, or I might be substituted for an H. We assume, of course, that the H can be retained in any or all of the available substitution positions. Suppose that we permit any one of k substituents at each of the six positions. Then k^6 patterns are initially possible, that is, $|\Omega| = k^6$. The symmetries are the elements of the dihedral group D_6 of order 12. We write $D_6 = \langle r, c\rangle$, where r is a rotation through $\pi/3$ and c is a reflection in an axis through two opposite vertices. The reader may check that r^2c and r^4c are also reflections about such an axis and that rc, r^3c, and r^5c are reflections about an axis drawn through the midpoints of two opposite sides of the hexagon. (See exercise 20.6.) Now the identity leaves fixed all k^6 configurations. The rotation r leaves fixed only those patterns having the same substituent attached to all six C's and thus $\chi(r) = k$; similarly, $\chi(r^5) = k$. The rotation r^2 preserves those configurations shown in figure 15(a); here one has k free choices for a and k free choices for b; thus $\chi(r^2) = k^2$, and similarly, $\chi(r^4) = k^2$. The 180° rotation r^3 preserves the patterns of figure 15(b); we have free choices for a, b, and c, and hence $\chi(r^3) = k^3$. From figure 15(c) we see that

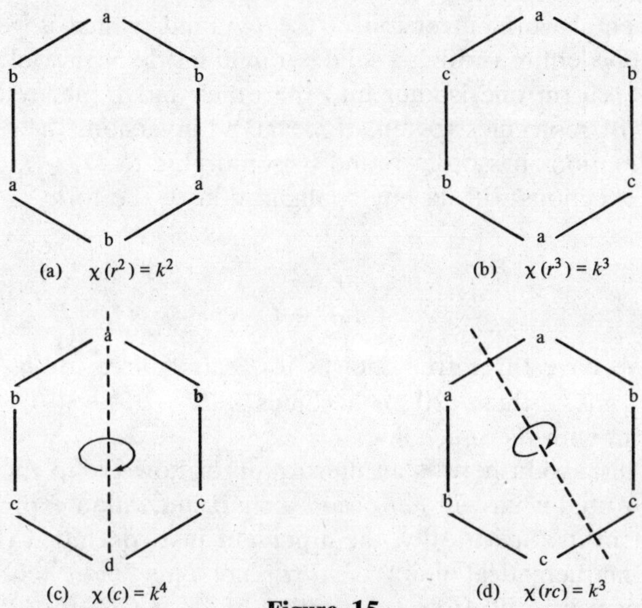

Figure 15

$\chi(c)=k^4$, and similarly for r^2c and r^4c. Finally, from figure 15(d), we have $\chi(rc)=\chi(r^3c)=\chi(r^5c)=k^3$. Substituting in (20.2), we obtain $\frac{1}{12}(k^6+2k+2k^2+k^3+3k^4+3k^3)$. We have proved:

(20.4) PROPOSITION. If any one of k different symbols may be placed at each vertex of a regular hexagon, then the number of distinct patterns formed (allowing for rotations and reflections) is

$$\frac{1}{12}(k^6+3k^4+4k^3+2k^2+2k). \tag{20.5}$$

In particular, for $k=2$, 3, 4, and 5, (20.5) equals 13, 92, 430, and 1505, respectively. For some applications in chemistry one speaks of only three "disubstituted benzenes" and ten "trisubstituted benzenes." Here the assumption is that only two or three of the hydrogens have been replaced and that the substituents are distinct. Under these constraints the result is easily arrived at by drawing the diagrams; no group theory is needed.

The use of the letter χ in the statement of the Burnside Counting Theorem suggests that the function χ is a group character. In fact, we may think of the points of Ω as the numbers 1, 2, $\ldots$, n and identify the natural basis vectors $e_1, e_2, \ldots, e_n$ with them as we did in section 15. Then, as in that section, for $g \in G$ take $T(g)$ to be the permutation matrix that carries e_i to e_j if and only if $i^g=j$, and let χ be the character associated with T. Then

$\chi(g)$ counts the number of points i of Ω for which $i^g = i$, as in (20.1). By comparing (20.2) with (14.1), we observe that the former is simply $\langle \chi, \zeta^{(1)} \rangle$, where $\zeta^{(1)}$ is the trivial character of G. Hence by (15.2), we have proved the following:

(20.6) THEOREM. Let G be a group acting on a point set Ω, and for each $g \in G$, let $\chi(g)$ be the number of points of Ω left fixed by g. Then χ is a character of G, and the multiplicity of the trivial character in χ is equal to the number of distinct orbits of Ω under G.

EXERCISES

20.1. Verify the conclusion of the Burnside Counting Theorem for S_4 regarded as a group acting on the eight vertices of the cube, and for A_4 as a group acting on the four vertices of the tetrahedron.

20.2. The pentaborane molecule B_5H_9 may be thought of as a square pyramid, as shown in figure 16. (The hydrogens shown on the sides

Figure 16

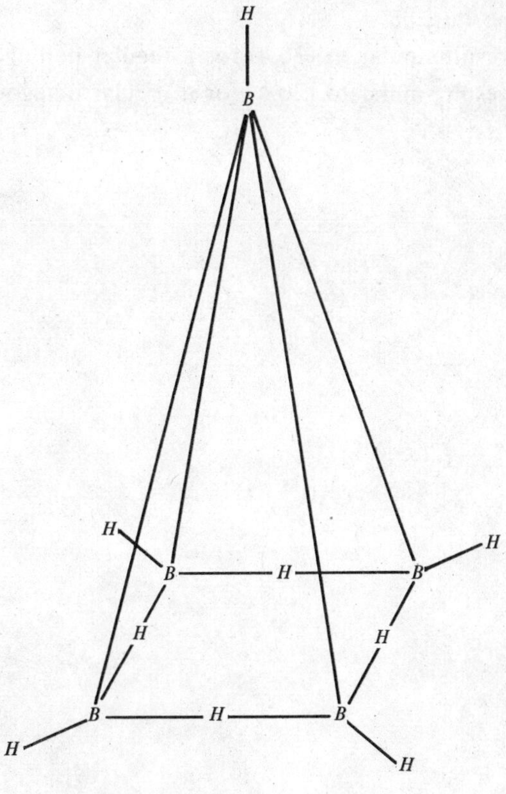

H
B
H
B
H
B
H
H
H
B
H
B
H
H

of the base are in fact below the plane of the square, but for our purposes the figure may be simplified as shown.) Suppose that we allow substitution of a halogen (F, Cl, Br, or I) or retention of the H at each of the four H-positions attached to the vertices of the base, and that we do *not* permit inversion. Show that 165 visually distinct configurations result.

20.3. Show that if inversion is permitted in exercise 20.2, the number of distinct configurations drops to 120.

20.4. Suppose that in exercise 20.2 we allow substitution of a halogen or retention of the H at each of the nine H-positions. How many distinct configurations may be formed?

20.5. The methane molecule CH_4 has the form of a tetrahedron with H at each vertex and C at the center of gravity. If substitution of a halogen or retention of the H is allowed at each vertex, how many distinct configurations may be formed?

20.6. Number the vertices of a regular hexagon from 1 to 6; let $r = (123456)$ and $c = (26)(35)$. Verify that r^2c and r^4c are also reflections in an axis drawn through two opposite vertices, and that each of rc, r^3c, and r^5c is a reflection in an axis joining the midpoints of two opposite sides.

20.7. Find a result similar to (20.4) for a regular pentagon.

20.8. Find a result similar to (20.4) for a regular octagon.

APPENDIX

Proofs of the Sylow Theorems

To complete the presentation of the material in section 6, here are proofs of the Sylow theorems and of propositions (6.2) and (6.3). Our approach emphasizes the concept of a group acting on a set and is derived from that of Helmut Wielandt [**9**]. For a more classical approach, see Dean [**5**], pp. 239–245, or Hall [**7**], pp. 44–46.

Before taking up the proof of the first Sylow theorem, let us make two observations:

(A.1) REMARK. Let V be the power set of G (the set of all subsets of G), and for $g \in G$ and $U \in V$, let

$$U^g = \{ug : u \in U\} \quad \text{for} \quad U \in V. \tag{A.2}$$

Then (A.2) defines an action of G on V. (This result was given in exercise 2.14; the proof is trivial.) Furthermore, since $|U^g| = |U|$ for any subset U of G (by the same proof as for (3.6)), if V is the collection of all subsets of G having a given order k, then (A.2) also defines an action of G on this V.

(A.3) REMARK. The number of distinct subsets of order m that can be made from a population of order n (usually referred to as "the combinations of n things taken m at a time") is given by the binomial coefficient

$$\binom{n}{m} = \frac{n!}{m!(n-m)!}.$$

Now we are ready to prove:

(6.1) SYLOW THEOREM I. Let p be a prime and $|G| = p^e q$, where $p \nmid q$. Then G contains a subgroup of order p^e, called a *Sylow p-subgroup.*

PROOF. We begin by taking V in (A.1) to be the collection of subsets of G having order p^e; *we do not assume* that any such subset is a subgroup! By (A.3), the number of such subsets is

$$\text{(A.4)} \qquad \binom{|G|}{p^e} = \frac{|G|(|G|-1)(|G|-2)\ldots(|G|-k)\ldots(|G|-(p^e-1))}{p^e \cdot \quad 1 \quad \cdot \quad 2 \cdot \ldots \quad \cdot \quad k \cdot \quad \cdots \quad \cdot \quad (p^e-1)}$$

Now we shall show that whenever a power of p appears in the numerator, it cancels one in the denominator. First, $|G|/p^e$ at the beginning of (A.4) reduces to q, which is not divisible by p. Now suppose that p^a divides $(|G|-k)$ in the numerator for some $a \geq 1$. If $a \geq e$, then p^e also divides $(|G|-k)$, and since p^e divides $|G|$, it also divides k, which contradicts the fact that $k \leq p^e - 1$. Hence $a < e$, which implies that p^a divides $|G|$ and consequently that p^a divides k, as was to be shown. Now on the one hand, any power of p in the numerator of (A.4) cancels one in the denominator, and on the other hand, it is well known that $\binom{n}{m}$ is a positive integer; hence p does not divide (A.4), which is $|V|$.

We let G act on V by right-multiplication, as in (A.1), and let $V_1, \ldots, V_n$ be the distinct orbits of V under this action. For each $i = 1, \ldots, n$, let $V_i = U_i^G$, that is, U_i is a typical element of the orbit V_i. Now

$$|V| = \sum_{i=1}^{n} |V_i|,$$

and if p divides *every* $|V_i|$, then p divides $|V|$, which is a contradiction. Hence there exists at least one i such that $p \nmid |V_i|$; suppose $i = 1$. Let S be the stabilizer of the subset U_1; we shall show that S is a Sylow p-subgroup of G. By (3.3), S is a subgroup, so we need prove only that it has order p^e.

If $g \in S$, then $U_1^g = U_1$, and if $u \in U_1$, then $ug \in U_1$. Hence

$$g \in u^{-1}U_1 = \{u^{-1}x : x \in U_1\}.$$

Therefore, $S \subseteq u^{-1}U_1$ and $|S| \leq |u^{-1}U_1| = |U_1| = p^e$. On the other hand, we took $|V_1|$ as not divisible by p, and by (3.11), $|V_1| = |G|/|S|$. But (3.11) also says that $|V_1|$ divides $|G| = p^e q$. Hence $|V_1|$ must divide q and so is $\leq q$; consequently $|S| = |G|/|V_1| \geq p^e$. From the two inequalities we conclude that $|S| = p^e$; this completes the proof of (6.1).

We turn next to the results on *p-groups*. In section 6 we defined a finite group to be a *p-group* if its order is a power of the given prime p. A more general definition, applicable also to infinite groups, is that a group G is a p-group if for each $g \in G$ there exists a nonnegative integer k such that the order of g is p^k. Using this newer definition, we prove first:

(A.5) PROPOSITION. Let G be finite; then G is a p-group if and only if there exists a nonnegative integer n such that $|G| = p^n$.

PROOF. Let G be a p-group; if $|G| = 1$, the conclusion follows with $n = 0$, so assume $|G| > 1$ and let q be *any* prime dividing $|G|$. We shall show that $q = p$. Let S be a Sylow p-subgroup of G having order q^m; if $x \in S$, then by exercise 3.10 the order of x is a divisor of q^m and hence a power of q, say q^k. But G is a p-group, so the order of x is a power of p; hence $p = q$, and p is the only prime dividing $|G|$. Conversely, if $|G| = p^n$, then since the order of an element of G must be a divisor of $|G|$, x has a power of p as its order. This completes the proof and reconciles the two definitions of p-group in the case of a finite group G.

Now we are ready to prove:

(6.2) PROPOSITION. The center of a finite p-group has order greater than 1.

PROOF. Let G act on itself by conjugation, and let

$$A_1 = a_1^G, \qquad A_2 = a_2^G, \qquad \ldots, \qquad A_k = a_k^G$$

be the distinct conjugate classes of G (the orbits of the set G under the action of conjugation). Now for any i, $|A_i| = 1$ if and only if $a_i \in C(G)$, since for arbitrary g,

$$a_i^g = a_i \quad \text{iff} \quad g^{-1} a_i g = a_i \quad \text{iff} \quad a_i g = g a_i.$$

Hence G is the set-theoretic union of $C(G)$ (which is nonempty since at least $1 \in C(G)$) and the union of those orbits A_i that contain more than one element. Now by proposition (A.5),

$$p^n = |G| = |C(G)| + \sum\{|A_i| : 1 \le i \le k, |A_i| \ne 1\}. \tag{A.6}$$

By (3.9) and (3.11), $|A_i|$ divides $|G|$ and so is a power of p. Thus p divides every $|A_i|$ that is not equal to 1. Since p divides every other term in (A.6), it must also divide $|C(G)|$ and hence $|C(G)| > 1$.

We can use (6.2) to prove (6.3) as follows:

(6.3) PROPOSITION. If g is a finite p-group and k divides $|G|$, then G contains a subgroup of order k.

PROOF. Let $|G|=p^n$; since k divides $|G|$, there exists a nonnegative integer $m \leq n$ such that $k=p^m$. If $m=n$, there is nothing to prove. We assume that $m<n$ and proceed by induction on m. If $m=0$, then $\{1\}$ is the required subgroup. Assuming $m \geq 1$, let us consider two cases.

Case 1: G is abelian. Let $x \neq 1$ in G, and let p^r be the order of x. If $r=n$, then $x^{p^{n-m}}$ generates a subgroup of order p^m, which is what we needed. Hence let $r<n$. If $r=m$, then $\langle x \rangle$ is the required subgroup, while if $r>m$, then $x^{p^{r-m}}$ generates a subgroup of order p^m. If $r<m$, then since G is abelian, $\langle x \rangle \trianglelefteq G$, and $G/\langle x \rangle$ is a group of order p^{n-r}. Now $m-r<m$, so by the induction hypothesis, $G/\langle x \rangle$ contains a subgroup $H/\langle x \rangle$ of order p^{m-r}, and H is a subgroup of order p^m in G. This completes the first case.

Case 2: G is nonabelian. By (6.2), there exists an integer s, $1 \leq s<n$, such that $|C(G)|=p^s$. If $m<s$, then the result follows by case 1 together with exercise 3.3. If $m=s$, then $C(G)$ is the required subgroup. If $m>s$, then $G/C(G)$ has order p^{n-s} and by the induction hypothesis contains a subgroup $H/C(G)$ of order p^{m-s}, whence H is a subgroup of G having order p^m. This completes the proof.

We now know that if p is a prime and if p^k divides $|G|$, then G has a subgroup of order p^k. We proceed to the proofs of the other Sylow theorems.

(6.5) SYLOW THEOREM II. Let $|G|=p^e q$ with p prime and $p \nmid q$. Let $H \leq G$ with $|H|=p^k$, and let P be *any* Sylow p-subgroup of G. Then there exists an element $g \in G$ such that $H \leq g^{-1}Pg$.

PROOF. Let S be the Sylow p-subgroup obtained in the proof of the first Sylow theorem; S is the stabilizer of a subset U_1 of order p^e under the action (A.2) of right multiplication. We shall establish the result first under the assumption that $P=S$ and then generalize to the case of $P \neq S$.

Case 1: $P=S$. By restriction of the action (A.2) to the subgroup H, it follows that H acts on the set $V_1=U_1^G$. However, we shall expect V_1 to have more than one orbit under H since only those elements of G that are in H are available to carry subsets in V_1 to one another. Let $X_1, \ldots, X_m$ be the distinct orbits of V_1 under H, with $X_i=Y_i^H$ for each $i=1, \ldots, m$. Note that each $Y_i \in V_1$. Then by (3.11), $|X_i|=|H|/|H_{x_i}|$, and since $|H|$ is a power of p, there exist nonnegative integers $c_1, \ldots, c_m$ such that $|X_i|=p^{c_i}$

for each i. But $|V_1| = \sum_{i=1}^{m} |X_i|$ and p does not divide $|V_1|$; hence there exists at least one i such that $p \nmid |X_i|$, and we may assume that $p \nmid |X_1|$. Now $|X_1|$ is a power of p and is not divisible by p; this can occur only if $|X_1| = p^0 = 1$. Hence H is contained in the stabilizer of Y_1. Since $Y_1 \in V_1$, there exists $g \in G$ such that $Y_1 = U_1^g$. Since S is the stabilizer of U_1, we conclude by (5.4) that H is contained in $g^{-1}Sg$. This completes the argument for $P = S$.

Case 2: $P \neq S$. By the case just completed (with $k = e$) we know that there exists $a \in G$ such that $P \leq a^{-1}Sa$. Now $|P| = |a^{-1}Sa|$, so $P = a^{-1}Sa$. Also by the first case, there exists $b \in G$ such that $H \leq b^{-1}Sb$. Let $g = a^{-1}b$; then

$$g^{-1}Pg = (b^{-1}a)(a^{-1}Sa)(a^{-1}b) = b^{-1}Sb,$$

and hence $H \leq g^{-1}Pg$. This completes the proof of (6.5).

(6.8) SYLOW THEOREM III. Let $|G| = p^e q$ with p prime and $p \nmid q$. Then the number s_p of distinct Sylow p-subgroups of G is a divisor of q, and p divides $(s_p - 1)$.

PROOF. First, (6.5) says that any two Sylow p-subgroups are conjugate and hence, for the given prime p, the set of Sylow p-subgroups of G has only one orbit under the action of conjugation; thus s_p is the order of that one orbit. If P is a Sylow p-subgroup of G and N is the normalizer of P, then $s_p = [G:N]$ by (3.11). Now $P \leq N$, so $|P|$ divides $|N|$ and hence by (3.10),

$$|G| = |N|[G:N] = |P|[G:P] \text{ and}$$
$$s_p = [G:N] \quad \text{divides} \quad [G:P] = q.$$

This completes the first part of the proof.

For the second part, we must show that there exists a nonnegative integer k such that $s_p = 1 + kp$. Let $\Omega = \{P_1, \ldots, P_{s_p}\}$ be the set of all Sylow p-subgroups of G (for our fixed p), and let P_1 act on the set Ω by conjugation. For each $i = 1, \ldots, s_p$, let N_i be the normalizer in P_1 of P_i, that is, N_i consists of those elements g of P_1 for which $g^{-1}P_ig = P_i$. Now Ω is the union of its distinct orbits $\Omega_1, \ldots, \Omega_r$ under P_1; let $k_j = |\Omega_j|$ for $j = 1, \ldots, r$. If $P_i \in \Omega_j$, then $k_j = [P_1:N_i]$. By the closure property in a group, $N_1 = P_1$ and we may take $\Omega_1 = \{P_1\}$ with $k_1 = 1$. We show now that $k_j > 1$ for $2 \leq j \leq r$. If $k_j = 1$, then $\Omega_j = \{P_i\}$ for some i, and $P_1 = N_i$. Hence $P_1P_i = P_iP_1$ and by (3.5), $P_1P_i \leq G$. Moreover, since P_1 and P_i are both contained in $N_G P_i$, we have $P_i \trianglelefteq P_1 P_i$. By (5.10),

$$P_1 P_i / P_i \cong P_1 / (P_1 \cap P_i);$$

hence by (3.10), $|P_1 P_i|$ is the product of $|P_i|$ and $|P_1/(P_1 \cap P_i)|$, which makes $|P_1 P_i|$ a power of p. But since P_i is a Sylow p-subgroup of G, $|P_1 P_i| = |P_i|$; hence $P_1 = P_i$ and $i = 1$, whence $j = 1$. This establishes the fact that $k_j > 1$ for $2 \leq j \leq r$. Now $k_j = [P_1 : N_i]$ is a divisor of $|P_1|$ and hence a power of p. Thus p divides the sum $k_2 + \ldots + k_r$, and we conclude that there is an integer $k \geq 0$ such that

$$\begin{aligned} s_p &= k_1 + (k_2 + \ldots + k_r) \\ &= 1 + kp. \end{aligned}$$

Bibliography

REFERENCES

[**1**] Bolker, Ethan D. "The Spinor Spanner." *American Mathematical Monthly* 80 (1973): 977–984.

[**2**] Burnside, W. *Theory of Groups of Finite Order.* 2d ed. New York: Dover, 1911 (reprinted 1955).

[**3**] Curtis, Charles W. *Linear Algebra: An Introductory Approach.* 3d ed. Boston: Allyn and Bacon, 1974.

[**4**] Curtis, Charles W., and Irving Reiner. *Representation Theory of Finite Groups and Associative Algebras.* New York: Interscience, 1962.

[**5**] Dean, Richard A. *Elements of Abstract Algebra.* New York: Wiley, 1966.

[**6**] Fraleigh, John B. *A First Course in Abstract Algebra.* 2d ed. Reading, Mass.: Addison-Wesley, 1976.

[**7**] Hall, Marshall, Jr. *The Theory of Groups.* New York: Macmillan, 1959.

[**8**] Scott, W. R. *Group Theory.* Englewood Cliffs, N.J.: Prentice-Hall, 1964.

[**9**] Wielandt, Helmut. "Zum Satz von Sylow." *Mathematische Zeitschrift* 60 (1954): 407–408.

SUGGESTIONS FOR FURTHER READING

Benson, C. T., and L. C. Grove. *Finite Reflection Groups.* Tarrytown-on-Hudson, N.Y.: Bogden & Quigley (Prindle, Weber & Schmidt), 1971.

Buerger, M. J. *Elementary Crystallography.* New York: Wiley, 1956.

Burrow, Martin. *Representation Theory of Finite Groups.* New York: Academic Press, 1965.

Cornwell, J. F. *Group Theory and Electronic Energy Bands in Solids.* Amsterdam: North-Holland, 1969.

Cotton, F. Albert. *Chemical Applications of Group Theory.* 2d ed. New York: Wiley-Interscience, 1971.

Dornhoff, Larry. *Group Representation Theory.* 2 vols. New York: M. Dekker, 1971, 1972.

Falicov, L. M. *Group Theory and Its Physical Applications.* Chicago: University of Chicago Press, 1965.

Feit, Walter. *Characters of Finite Groups.* New York: W. A. Benjamin, 1967.

Hall, G. G. *Applied Group Theory.* London: Longmans, 1967.

Heine, Volker. *Group Theory in Quantum Mechanics: An Introduction to Its Present Usage.* London: Pergamon, 1960.

Hermann, Robert. *Lie Groups for Physicists.* New York: W. A. Benjamin, 1966.

Higman, Bryan. *Applied Group-Theoretic and Matrix Methods.* Oxford: Clarendon Press, 1955.

Hochstrasser, Robin M. *Molecular Aspects of Symmetry.* New York: W. A. Benjamin, 1966.

Hollingsworth, Charles A. *Vectors, Matrices, and Group Theory for Scientists and Engineers.* New York: McGraw-Hill, 1967.

Jaffé, H. H., and Milton Orchin. *Symmetry in Chemistry.* New York: Wiley, 1965.

Kettle, S. F. A. "Ligand Group Orbitals of Octahedral Complexes." *J. Chemical Education* 43 (1966): 21–26.

Leech, J. W., and D. J. Newman. *How to Use Groups.* London: Methuen, 1969.

Lichtenberg, Don Bernett. *Unitary Symmetry and Elementary Particles.* New York: Academic Press, 1970.

Lipkin, Harry J. *Lie Groups for Pedestrians.* Amsterdam: North-Holland, 1966.

Loebl, Ernest M., ed. *Group Theory and Its Applications.* 2 vols. New York: Academic Press, 1968, 1971.

Lomont, J. S. *Applications of Finite Groups.* New York: Academic Press, 1959.

Mariot, L. *Group Theory and Solid State Physics.* Englewood Cliffs, N.J.: Prentice-Hall, 1962.

McWeeny, R. *Symmetry: An Introduction to Group Theory and Its Applications.* Oxford: Pergamon, 1963.

Meijer, Paul Herman Ernst. *Group Theory and Solid State Physics.* New York: Gordon and Breach, 1964.

Miller, Willard, Jr. *Symmetry Groups and Their Applications.* New York: Academic Press, 1972.

Nussbaum, Allen. *Applied Group Theory for Chemists, Physicists, and Engineers.* Englewood Cliffs, N.J.: Prentice-Hall, 1971.

Park, David. "Resource Letter SP-1 on Symmetry in Physics." *American J. Physics* 36 (1968): 1–8.

Petrashen, M. I., and E. D. Trifonov. *Group Theory in Quantum Mechanics.* Translated by S. Chomet, edited and revised by J. L. Martin. Cambridge, Mass.: M.I.T. Press, 1969.

Racah, Giulio. "Group Theory and Spectroscopy." *Ergebnisse der exacten Naturwissenschaften* 37 (1965): 28–84.

Robinson, G. de B. *Representation Theory of the Symmetric Group.* Toronto: University of Toronto Press, 1961.

Serre, Jean-Pierre. *Représentations Linéaires des Groupes Finis.* Paris: Hermann, 1967.

Schonland, D. S. *Molecular Symmetry: An Introduction to Group Theory and Its Uses in Chemistry.* New York: Van Nostrand, 1965.

Weyl, Hermann. *The Theory of Groups and Quantum Mechanics.* Translated by H. P. Robertson. New York: Dover, 1950.

Wigner, Eugene Paul. *Group Theory and Its Application to the Quantum Mechanics of Atomic Spectra.* Expanded and improved ed. Translated by J. J. Griffin. New York: Academic Press, 1959.

Zeldin, Martel. "An Introduction to Molecular Symmetry and Symmetry Point Groups." *J. Chemical Education* 43 (1966): 17–20.

Index

Index of Symbols

The page reference is to the first occurrence of the symbol.